Springer Theses

Recognizing Outstanding Ph.D. Research

Aims and Scope

The series "Springer Theses" brings together a selection of the very best Ph.D. theses from around the world and across the physical sciences. Nominated and endorsed by two recognized specialists, each published volume has been selected for its scientific excellence and the high impact of its contents for the pertinent field of research. For greater accessibility to non-specialists, the published versions include an extended introduction, as well as a foreword by the student's supervisor explaining the special relevance of the work for the field. As a whole, the series will provide a valuable resource both for newcomers to the research fields described, and for other scientists seeking detailed background information on special questions. Finally, it provides an accredited documentation of the valuable contributions made by today's younger generation of scientists.

Theses are accepted into the series by invited nomination only and must fulfill all of the following criteria

- They must be written in good English.
- The topic should fall within the confines of Chemistry, Physics, Earth Sciences, Engineering and related interdisciplinary fields such as Materials, Nanoscience, Chemical Engineering, Complex Systems and Biophysics.
- The work reported in the thesis must represent a significant scientific advance.
- If the thesis includes previously published material, permission to reproduce this must be gained from the respective copyright holder.
- They must have been examined and passed during the 12 months prior to nomination.
- Each thesis should include a foreword by the supervisor outlining the significance of its content.
- The theses should have a clearly defined structure including an introduction accessible to scientists not expert in that particular field.

More information about this series at http://www.springer.com/series/8790

Tsukasa Sawato

Synthesis of Optically Active Oxymethylenehelicene Oligomers and Self-assembly Phenomena at a Liquid–Solid Interface

Doctoral Thesis accepted by
Tohoku University, Sendai, Japan

Author
Dr. Tsukasa Sawato
Tohoku University
Sendai, Japan

Supervisor
Prof. Masahiko Yamaguchi
Department of Organic Chemistry,
Graduate School of Pharmaceutical Sciences
Tohoku University
Sendai, Japan

ISSN 2190-5053 ISSN 2190-5061 (electronic)
Springer Theses
ISBN 978-981-15-3194-1 ISBN 978-981-15-3192-7 (eBook)
https://doi.org/10.1007/978-981-15-3192-7

This Springer imprint is published by the registered company Springer Nature Singapore Pte Ltd.
The registered company address is: 152 Beach Road, #21-01/04 Gateway East, Singapore 189721, Singapore

Supervisor's Foreword

It is my great pleasure to introduce the work of Dr. Tsukasa Sawato for publication in the Springer Theses series as an outstanding original work from one of the world's top universities. Dr. Sawato joined my laboratory in Tohoku University in October 2012 as an undergraduate student. In April 2014, he entered the Graduate School of Pharmaceutical Sciences at Tohoku University and started his doctoral study with me and Dr. Masanori Shigeno. He was accepted as a Japan Society for the Promotion of Science (JSPS) doctoral fellow in April 2018, and he received his doctor's degree in March 2019.

Chemical reactions are ubiquitous in nature and living things, and their understanding and control are important subjects of chemistry. Among them, covalent and noncovalent bond rearrangements, conformation changes, and aggregations of organic molecules at liquid–solid interfaces are important phenomena and exhibit different properties from those in solution. The differences can be ascribed to factors including interactions of molecules with the solid surface, a high concentration of molecules on the solid surface, and the confined conformation of molecules on the solid surface. Little, however, is known about the chemical reactivities of molecules at the interface compared with those in solution; this may be due to the extremely low concentration of molecules at the interface. Dr. Sawato tackled this subject employing helix-forming synthetic oligomers and has found and analyzed various phenomena, which are characteristics of the interfaces. The notable results are highlighted in this thesis. As shown in his additional publication list, Dr. Sawato has also been involved in the related studies on dynamic properties of several helix-forming oligomers.

Knowing his passion, executive ability, leadership, and communication ability, I am quite convinced that Dr. Sawato will become a leading chemist in our scientific society.

Sendai, Japan
January 2020

Prof. Masahiko Yamaguchi

Parts of this thesis have been published in the following journal articles:

1. Fibril Film Formation of Pseudoenantiomeric Oxymethylenehelicene Oligomers at the Liquid–Solid Interface: Structural Changes, Aggregation, and Discontinuous Heterogeneous Nucleation
Masanori Shigeno, Tsukasa Sawato, and Masahiko Yamaguchi, *Chem. Eur. J.* **2015**, *21*, 17676–17682.

2. Mechanical Stirring Induces Heteroaggregate Formation and Self-assembly of Pseudoenantiomeric Oxymethylene Helicene Oligomers in Solution
Tsukasa Sawato, Nozomi Saito, Masanori Shigeno, and Masahiko Yamaguchi, *ChemistrySelect* **2017**, *2*, 2205–2211.

3. Proximate stochastic chiral symmetry breaking is mechanically tunable: formation of enantiomeric hetero-double-helices and aggregates from racemic oxymethylenehelicene oligomers
Tsukasa Sawato, Nozomi Saito, and Masahiko Yamaguchi, *Phys. Chem. Chem. Phys.* **2019**, *21*, 25406–25414.

Related Papers

Spatially Heterogeneous Nature of Self-Catalytic Reaction in Hetero-Double Helix Formation of Helicene Oligomers
Yo Kushida, Tsukasa Sawato, Nozomi Saito, Masanori Shigeno, Hiroshi Satozono, and Masahiko Yamaguchi, *ChemPhysChem* **2016**, *17*, 3283–3288.

Deterministic and Stochastic Chiral Symmetry Breaking Exhibited by Racemic Aminomethylenehelicene Oligomers
Yo Kushida, Tsukasa Sawato, Masanori Shigeno, Nozomi Saito, and Masahiko Yamaguchi, *Chem. Eur. J.* **2017**, *23*, 327–333.

Chemical braking exhibited by ethynylhelicene (*M*)-nonamer in solution: Competitive reaction system of self-catalysis to form double-helix and approach towards equilibrium to form random-coil
Tsukasa Sawato, Atsushi Yagi, Mieko Arisawa, and Masahiko Yamaguchi, *Tetrahedron* **2017**, *73*, 2801–2805.

Pendant-Type Helicene Oligomers with *p*-Phenylene Ethynylene Main Chains: Synthesis, Reversible Formation of Ladderlike Bimolecular Aggregates, and Control of Intramolecular and Intermolecular Aggregation
Nozomi Saito, Yutaro Kondo, Tsukasa Sawato, Masanori Shigeno, Ryo Amemiya, and Masahiko Yamaguchi, *J. Org. Chem.* **2017**, *82*, 8389–8406.

Optically active iodohelicene derivatives exhibit histamine N-methyl transferase inhibitory activity
Wataru Ichinose, Tsukasa Sawato, Haruna Kitano, Yasuhiro Shinozaki, Mieko Arisawa, Nozomi Saito, Takeo Yoshikawa, Masahiko Yamaguchi, *J. Antibiot.* **2019**, *72*, 476-481.

Chemical CD oscillation and chemical resonance phenomena in a competitive self-catalytic reaction system: a single temperature oscillation induces CD oscillations twice
Tsukasa Sawato, Yasuhiro Shinozaki, Nozomi Saito and Masahiko Yamaguchi, *Chem. Sci.* **2019**, *10*, 1735–1740.

Chemical Systems Involving Two Competitive Self-Catalytic Reactions
Tsukasa Sawato, Nozomi Saito, and Masahiko Yamaguchi, *ACS Omega* **2019**, *4*, 5879–5899.

Formation and dissociation of synthetic heterodouble-helix complex in aqueous solutions: significant effect of water content on dynamics of structural change
Tsukasa Sawato, Ryosuke Yuzawa, Higashi Kobayashi, Nozomi Saito, and Masahiko Yamaguchi, *RSC Adv.* **2019**, *9*, 29456–29462.

Acknowledgements

I express my sincere and wholehearted appreciation to Prof. Dr. Masahiko Yamaguchi (Graduate School of Pharmaceutical Sciences, Tohoku University) for his constructive discussions, warm encouragement, and guidance in the ways of thinking as a chemist.

I am profoundly grateful for the kind support and detailed, careful guidance given to me by Dr. Masanori Shigeno (Graduate School of Pharmaceutical Sciences, Tohoku University).

I also would like to express my gratitude to Dr. Mieko Arisawa (Graduate School of Pharmaceutical Sciences, Tohoku University) and Dr. Nozomi Saito (Graduate School of Pharmaceutical Sciences, Tohoku University) for valuable comments and kind encouragement.

I gratefully acknowledge Prof. Dr. Takayuki Doi (Graduate School of Pharmaceutical Sciences, Tohoku University) and Prof. Dr. Yoshiharu Iwabuchi (Graduate School of Pharmaceutical Sciences, Tohoku University) for their thoughtful comments on the revision of my original manuscript.

In addition, I am grateful to Ms. Misa Sakuma, Ms. Ikuko Goto, and Ms. Aiko Abe for their official work and to all colleagues at the Yamaguchi Laboratory (Graduate School of Pharmaceutical Sciences, Tohoku University) for their cooperation, encouragement, and competitive spirit.

The Japan Society for the Promotion of Science (JSPS) is gratefully acknowledged for financial support and for the fellowship of young Japanese scientists that I received.

Finally, I am grateful to my parents, Akihiro and Naomi Sawato, and to all members of my family for their constant support and encouragement.

Contents

Chapter 1
Introduction

Abstract Chemical reactions and molecular structural changes of organic molecules at liquid–solid interfaces are important in biology and material sciences. Then, on the studies molecular structural change and aggregation of synthetic organic molecules at liquid–solid interfaces is interesting.

Keywords Liquid–solid interface · Helicene oligomer

1.1 Chemical Reactions at Liquid–Solid Interfaces

Chemical reactions, conformation changes, and aggregations of organic molecules at liquid–solid interfaces are important phenomena in nature and exhibit properties different from those in solution [1–3]. The differences can be ascribed to factors including interactions of molecules with the surface, high concentrations of organic molecules on the surface, and the confined conformation of molecules on the surface [4, 5]. Little, however, is known about the chemical reactivities of organic molecules on the solid surface compared with those in solution; this may be due to the extremely low concentration of molecules on the surface.

Chemical reactions at liquid–solid interfaces exhibit properties different from those in bulk solution. In bulk solution, molecules constantly move around owing to thermal motion and are three-dimensionally dispersed in a uniform state. In contrast, this uniformity is lost at liquid–solid interfaces. Molecules in solution collide with a solid surface and remain on the solid surface as adsorbed species. The process by which molecules are adsorbed on a solid surface through chemical bonds is called chemical adsorption, whereas that by which molecules are adsorbed by van der Waals forces is called physical adsorption. The adsorbed molecules diffuse two-dimensionally on a solid surface, because the energy required for diffusion is generally smaller than that required for adsorption. The density of molecules on solid surfaces then becomes higher than that in solution, and chemical reactivity can be changed. The desorption of molecules back into bulk solution also occurs, which is an important process for heterogeneous catalysis. Note also that the degrees of freedom of molecular conformations decrease at the interfaces owing to the interactions of molecules with solid surfaces. For example, biomacromolecules such as

T. Sawato, *Synthesis of Optically Active Oxymethylenehelicene Oligomers and Self-assembly Phenomena at a Liquid–Solid Interface*, Springer Theses,
https://doi.org/10.1007/978-981-15-3192-7_1

proteins and nucleic acids can change their conformation or state of aggregation at cell surfaces, and important biological phenomena can take place. Because of these properties, studies that may lead to the clarification and control of chemical reactions at liquid–solid interfaces are of interest.

It is, however, extremely difficult to analyze chemical reactions of molecules at liquid–solid interfaces, at least partly because of the presence of very minute amounts of molecules at the interfaces. Various methods are required for the study such as the use of (1) compounds suitable for analysis at the interfaces, (2) suitable chemical reactions, and (3) suitable analytical methods.

In this work, I utilized several methods to analyze chemical reactions at liquid–solid interfaces. I used helicene oligomers developed in our laboratory, which can exhibit an extremely strong Cotton effect, enabling the analysis of very small amounts of molecules at liquid–solid interfaces. In addition, various structurally related oligomeric compounds can be used to compare their properties. For the chemical reaction, the interconversion between hetero-double helices and random coils was used, which induces large changes in the Cotton effect and is extremely sensitive to environmental changes. For the analytical method, circular dichroism (CD) spectroscopy was used, which can sensitively analyze the surface structures of oligomers. Other spectroscopic methods including nuclear magnetic resonance (NMR), infrared absorption spectrometry (IR), ultraviolet–visible absorption spectroscopy (UV–vis), and fluorescence spectroscopies were also used. Microscopy methods are useful to probe molecules on surfaces, and include optical microscopy, polarized microscopy, fluorescence microscopy, atomic force microscope (AFM), and scanning electron microscope (SEM). Aggregation in solution was examined in the solution phase by dynamic light scattering (DLS) and vapor pressure osmometry (VPO), the results of which were compared with those observed at the interfaces.

Two types of external stimulation are used in this type of study to induce hetero-double-helix formation at the interfaces. One is making oligomers come in contact with solid surfaces, and different surface materials may be used to modify chemical reactions. The other type of external stimulation is mechanical stirring with a magnetic stirrer, which is routinely performed in laboratories. Mechanical stirring causes friction, resulting in the generation of hot domains that can locally and temporarily reach several thousand Kelvin. Although the heat generated generally diffuses rapidly, helicene oligomers can respond to this mechanical stimulation at liquid–solid interfaces.

1.2 Double-Helix/Random-Coil Transition of Helicene Oligomers

We previously observed structural changes of helicene oligomers in dilute solution between random coils at high temperatures and homo-double helices at low temperatures [6–9]. Several helicene oligomers with different two-atom linkers, including acetylene [10–13], amides [14, 15], sulfonamide [16, 17], and aminomethylene [18], exhibited structural changes. It was also reported that 1:1 mixtures of pseudoenantiomeric ethynylhelicene [19–21] and aminomethylenehelicene [22] oligomers, which are enantiomeric oligomers with different helicene numbers, formed hetero-double helices; and the former self-assembled to form fibrils and gels. The helicene oligomers exhibit a novel structure and reactivity, which can be systematically varied [6–9, 19–21].

1.3 Outline of This Work

In this study, two-atom-linker helicene oligomers, namely, oxymethylenehelicene oligomers, were developed, which are the oxygen derivatives of aminomethylenehelicene oligomers (Fig. 1.1). These new oligomers also undergo interconversion between double helices and random coils, a process that may be used to examine chemical reactions at liquid–solid interfaces (Chap. 2).

The helicene oligomers were synthesized by ether formation between phenols and helicenylmethyl bromides. A pseudoenantiomeric mixture of (*P*)-Ox-**5**/(*M*)-Ox-**6** formed hetero-double helices at 5 °C at liquid–solid interfaces, which then formed fibril films. This is a novel self-assembly of nonpeptide synthetic compounds on a solid surface, which exhibit discontinuous heterogeneous nucleation. The hetero-double helices self-assemble to form particles that are 50 nm in diameter at the liquid–solid interfaces; when the particles cease their growth, fibrils start to form from the particles (Fig. 1.2) (Chap. 3).

(*P*)-Ox-**5**/(*M*)-Ox-**6** also formed hetero-double helices in solution, which then self-assembled at 25 °C as induced by mechanical stirring. Thus, different types of molecular self-assembly occurred under different conditions: Contact with solid surfaces without mechanical stirring formed hetero-double helices and fibril films on the surfaces at 5 °C; mechanical stirring formed hetero-double helices and self-assembled materials in solution at 25 °C. It was noted that the hetero-double helies and self-assembled materials yielded CD spectra indicating enantiomeric structures of the hetero-double helices (Chap. 4).

Racemic mixtures of hexamers (*P*)-Ox-**6** and (*M*)-Ox-**6** showed proximate stochastic chiral symmetry breaking when stirred mechanically. This behavior

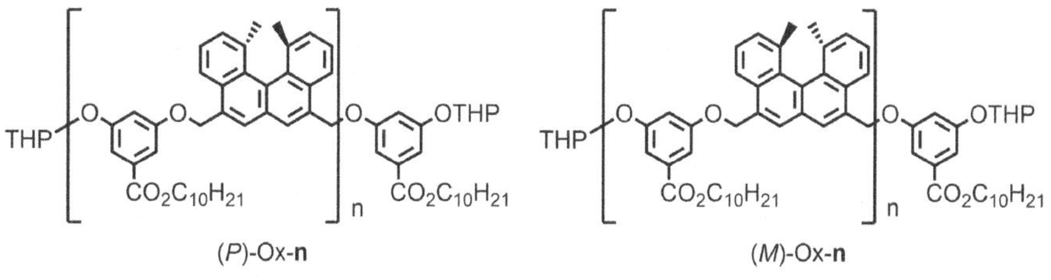

THP O O O OTHP $CO_2C_{10}H_{21}$ n $CO_2C_{10}H_{21}$ (*P*)-Ox-**n**

THP O O O OTHP $CO_2C_{10}H_{21}$ n $CO_2C_{10}H_{21}$ (*M*)-Ox-**n**

Fig. 1.1 Chemical structures of oxymethylenehelicene oligomers

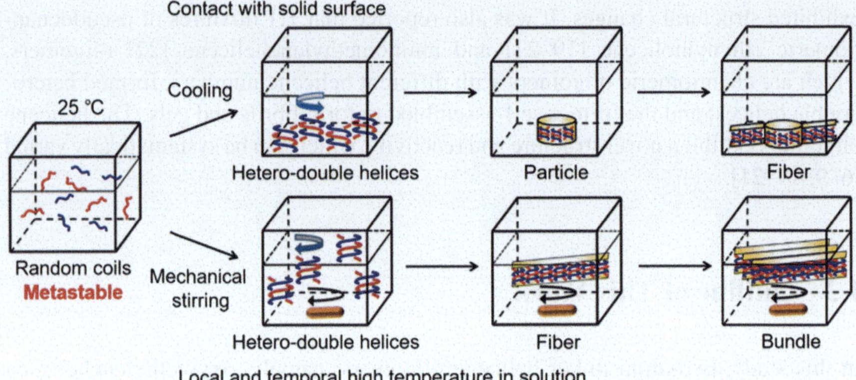

Fig. 1.2 Different self-assemblies of (*P*)-Ox-**5**/(*M*)-Ox-**6** depending on perturbations

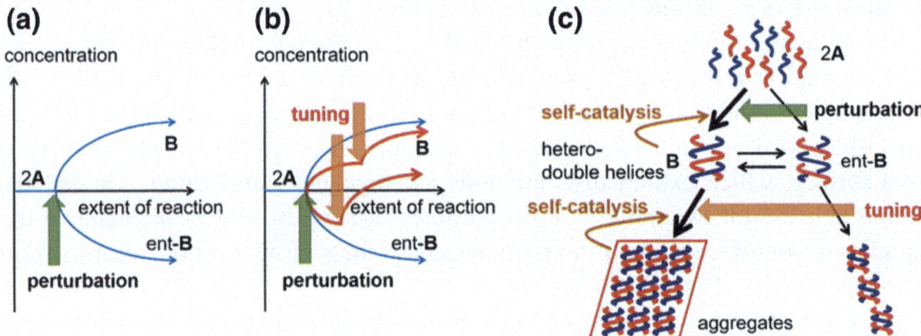

Fig. 1.3 **a** Schematic illustration of the process of chiral symmetry breaking in the reaction of random-coils 2**A** to form enantiomeric hetero-double helix **B** or ent-**B**, expressed by a bifurcation model with external perturbation. **b** Model with mechanical tuning of the process. **c** Structural aspects of chiral symmetry breaking in this study

implies chiral symmetry breaking very close to stochastic chiral symmetry breaking with a slight deviation of symmetry (Fig. 1.3). It was also noted that mechanical perturbations can stop and restart chiral symmetry breaking, and the enantiomeric products **B** and ent-**B** can be switched (Chap. 5).

Described in this thesis are (1) hetero-double-helix formation and fibril film formation on a solid surface as a result of the catalytic effect of liquid–solid interfaces, and (2) the hetero-double helices are generated at the liquid–solid interfaces and then diffuses into a solution to form a self-assembled material resulting from mechanical stirring. Both are novel chemical reactions unique to liquid–solid interfaces.

References

1. Somorjai GA, Li Y (eds) (2010) Introduction to surface chemistry and catalysis, 2nd edn. Wiley, Hoboken
2. Mali KS, Adisoejoso J, Ghijsens E, Cat ID, de Feyter S (2012) Exploring the Complexity of Supramolecular Interactions for Patterning at the Liquid–Solid Interface. Acc Chem Res 45:1309
3. Hecht S, Huc I (eds) (2007) Foldamers: structure, properties, and applications. Wiley-VCH, Weinheim
4. Cosgrove T (ed) (2010) Colloid science: principles, methods and applications, 2nd edn. Wiley, Chichester
5. Holmberg K (2002) Handbook of applied surface and colloid chemistry. Wiley, Chichester
6. Amemiya R, Yamaguchi M (2008) Chiral recognition in noncovalent bonding interactions between helicenes: right-handed helix favors right-handed helix over left-handed helix. Org Biomol Chem 6:26
7. Amemiya R, Yamaguchi M (2008) Synthesis and structure of built-up organic macromolecules containing helicene. Chem Rec 8:116
8. Yamaguchi M, Shigeno M, Saito N, Yamamoto K (2014) Synthesis, double-helix formation, and higher-assembly formation of chiral polycyclic aromatic compounds: conceptual development of polyketide aldol synthesis. Chem Rec 14:15
9. Shigeno M, Kushida Y, Yamaguchi M (2015) Energy Aspects of Thermal Molecular Switching: Molecular Thermal Hysteresis of Helicene Oligomers. ChemPhysChem 16:2076
10. Saito N, Terakawa R, Shigeno M, Amemiya R, Yamaguchi M (2011) Side Chain Effect on the Double Helix Formation of Ethynylhelicene Oligomers. J Org Chem 76:4841
11. Ichinose W, Shigeno M, Yamaguchi M (2012) Multiple States of Dimeric Aggregates Formed by (Amido–ethynyl)helicene Bidomain Compound and (Amido–ethynyl–amido)helicene Tridomain Compound. Chem Eur J 18:12644
12. Ichinose W, Ito J, Yamaguchi M (2012) Heteroaggregation between Isomeric Amido-ethynyl-amidohelicene Tridomain Oligomers. J Org Chem 77:10655
13. Miyagawa M, Yagi A, Shigeno M, Yamaguchi M (2014) Equilibrium crossing exhibited by an ethynylhelicene (*M*)-nonamer during random-coil-to-double-helix thermal transition in solution. Chem Commun 50:14447
14. Amemiya R, Ichinose W, Yamaguchi M (2010) Synthesis and Thermally Stable Helix-Dimer Formation of Amidohelicene Oligomers. Bull Chem Soc Jpn 83:809
15. Ichinose W, Miyagawa M, Ito J, Shigeno M, Amemiya R, Yamaguchi M (2011) Synthesis and duplex formation of the reverse amidohelicene tetramer. Tetrahedron 67:5477
16. Shigeno M, Kushida Y, Yamaguchi M (2013) Molecular Thermal Hysteresis in Helix-Dimer Formation of Sulfonamidohelicene Oligomers in Solution. Chem Eur J 19:10226
17. Shigeno M, Kushida Y, Yamaguchi M (2015) Self-catalysis in thermal hysteresis during random-coil to helix-dimer transition of the sulfonamidohelicene tetramer. Chem Commun 51:4040
18. Shigeno M, Sato M, Kushida Y, Yamaguchi M (2014) Aminomethylenehelicene Oligomers Possessing Flexible Two-Atom Linker Form a Stimuli-Responsive Double-Helix in Solution. Asian J Org Chem 3:797
19. Saito N, Shigeno M, Yamaguchi M (2012) Two-Component Fibers/Gels and Vesicles Formed from Hetero-Double-Helices of Pseudoenantiomeric Ethynylhelicene Oligomers with Branched Side Chains. Chem Eur J 18:8994
20. Yamamoto K, Oyamada N, Mizutani M, An Z, Saito N, Yamaguchi M, Kasuya M, Kurihara K (2012) Two Types of Two-Component Gels Formed from Pseudoenantiomeric Ethynylhelicene Oligomers. Langmuir 28:11939

21. Saito N, Kanie K, Matsubara M, Muramatsu A, Yamaguchi M (2015) Dynamic and Reversible Polymorphism of Self-Assembled Lyotropic Liquid Crystalline Systems Derived from Cyclic Bis(ethynylhelicene) Oligomers. J Am Chem Soc 137:6594
22. Shigeno M, Kushida Y, Yamaguchi M (2014) Heating/Cooling Stimulus Induces Three-State Molecular Switching of Pseudoenantiomeric Aminomethylenehelicene Oligomers: Reversible Nonequilibrium Thermodynamic Processes. J Am Chem Soc 136:7972

Chapter 2
Synthesis and Homo-double-helix Formation of Oxymethylenehelicene Oligomers

Abstract Oxymethylenehelicene (*P*)- and (*M*)-oligomers up to a nonamer were synthesized by a building block method. The oligomers formed homo-double-helix in trifluoromethylbenzene.

Keywords Helicene oligomer · Homo-double-helix

2.1 Synthesis

Oxymethylenehelicene (*P*)-oligomers containing odd numbers of helicenes, (*P*)-Ox-**n** (n = 3, 5, 7, 9), were synthesized by repeated ether formation using the building block (*P*)-**10** and removal of the THP protecting group (Scheme 2.2). Dibromohelicene (*P*)-**12**, obtained from a known helicenediol [1], was reacted with *m*-phenylene spacer **13** (1 equiv) in the presence of sodium hydride to yield the building block (*P*)-**10** (Scheme 2.1). The monomer (*P*)-Ox-**1** was synthesized by the coupling of (*P*)-**12** and **13** (2 equiv) in 88% yield. Removal of the THP group of (*P*)-Ox-**1** with *p*-toluenesulfonic acid (TsOH) and coupling with (*P*)-**10** afforded the trimer (*P*)-Ox-**3**. A series of coupling and deprotection reactions with (*P*)-Ox-**3** gave pentamer (*P*)-Ox-**5**, heptamer (*P*)-Ox-**7**, and nonamer (*P*)-Ox-**9** in high yields. (*M*)-Oligomers containing even numbers of helicenes, (*M*)-Ox-**n** (n = 4, 6), were synthesized by the same procedures starting from dimer (*M*)-Ox-**2** (Scheme 2.3).

T. Sawato, *Synthesis of Optically Active Oxymethylenehelicene Oligomers and Self-assembly Phenomena at a Liquid–Solid Interface*, Springer Theses,
https://doi.org/10.1007/978-981-15-3192-7_2

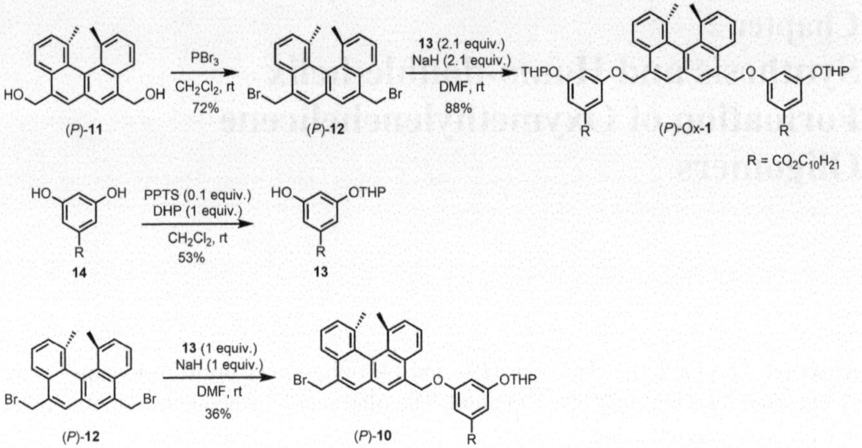

Scheme 2.1 Synthesis of (*P*)-building block

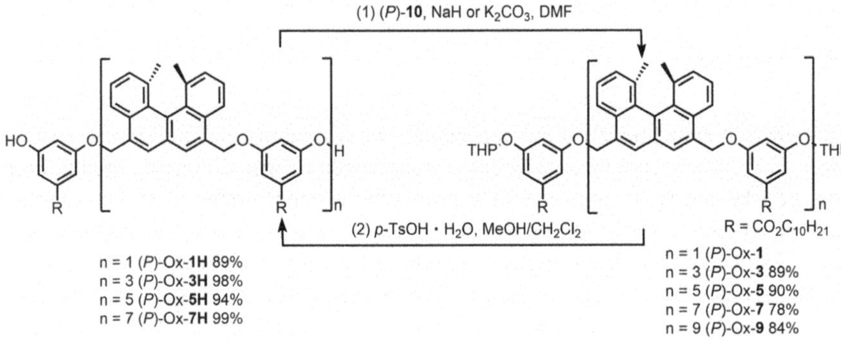

Scheme 2.2 Synthetic route of (*P*)-oxymethylenehelicene oligomers

2.2 Homo-double-helix Formation

Analogous to other helicene oligomers [2–10], the oxymethylene derivatives formed homo-double-helix in solution. In trifluoromethylbenzene, the structure of heptamer (*P*)-Ox-**7** (0.5 mM) changed with increasing temperature. At 80 °C, the ^{1}H NMR spectrum of (*P*)-Ox-**7** ([D_5]trifluoromethylbenzene; 2.0 mM) showed sharp signals consistent with the monomeric form (Fig. 2.1a). Vapor pressure osmometry (VPO) analysis (trifluoromethylbenzene, 80 °C) showed the monomeric nature of (*P*)-Ox-**7**: observed molecular weights: $(4.0 \pm 0.2) \times 10^3$ ($N = 0.89 \pm 0.04$) at 1.0 mM and $(4.4 \pm 0.2) \times 10^3$ ($N = 0.98 \pm 0.04$) at 2.0 mM; calculated monomer weight: 4486. When the solution was cooled to 25 °C, the NMR signals broadened (Fig. 2.1b). The Cotton effects at $\lambda = 300$ nm slightly shifted to $\lambda = 303$ nm with broadening upon cooling from 80 to 25 °C (Fig. 2.2). VPO analysis (trifluoromethylbenzene; 40 °C) showed dimeric homoaggregation between 4.0 and 6.0 mM (Fig. 2.3a). Dynamic

HO OH
R
14
(*M*)-**10**, NaH
DMF, rt
93%
THP O O O O THP
R R
(*M*)-Ox-**2**

(1) (*M*)-**10**, NaH or K_2CO_3, DMF
HO O O O H
R R n
THP O O O O THP
R R n
(2) *p*-TsOH · H_2O, MeOH/CH_2Cl_2
R = $CO_2C_{10}H_{21}$
n = 2 (*M*)-Ox-**2H** 95%
n = 4 (*M*)-Ox-**4H** 76%
n = 2 (*M*)-Ox-**2**
n = 4 (*M*)-Ox-**4** 81%
n = 6 (*M*)-Ox-**6** 80%

Scheme 2.3 Synthetic route of (*M*)-oxymethylenehelicene oligomers

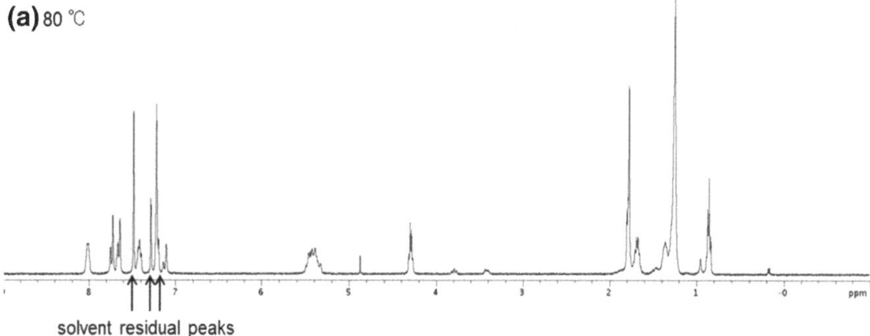

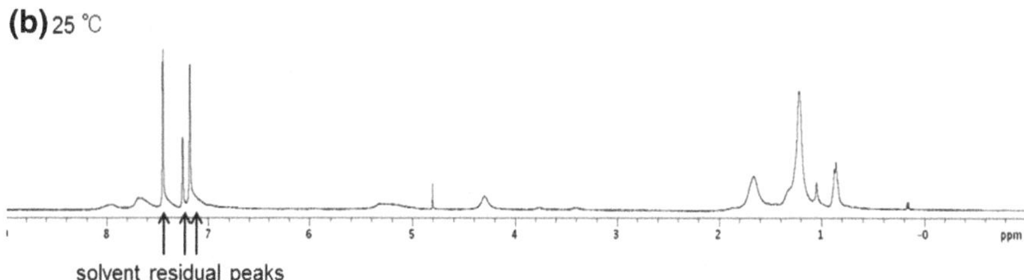

Fig. 2.1 ^{1}H NMR of (*P*)-Ox-**7** (n = 7). **a** in trifluoromethylbenzene-d_5 (2.0 mM, 80 °C). **b** in trifluoromethylbenzene-d_5 (2.0 mM, 25 °C)

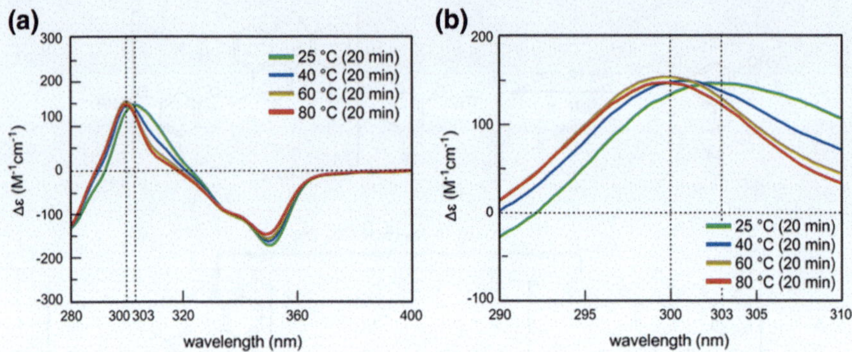

Fig. 2.2 **a** Circular dichroism (CD) spectra of (*P*)-Ox-**7** in trifluoromethylbenzene (0.5 mm) obtained at each temperature, after allowing the solution to stand for the time shown in parentheses, when the $\Delta\varepsilon$ ($\lambda = 303$ nm) change was less than 2 cm^{-1} M^{-1} within a 10 min period. **b** Expansion of (**a**)

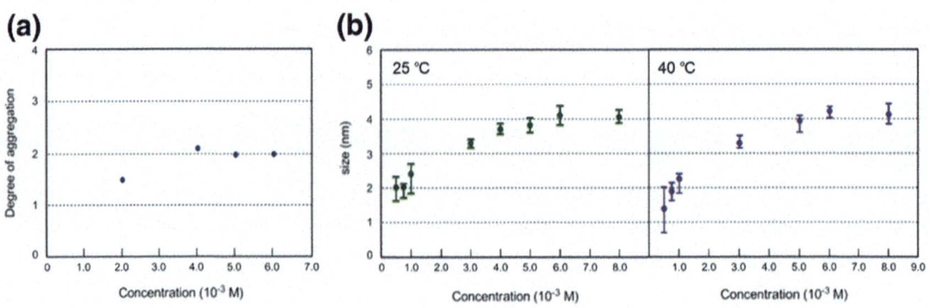

Fig. 2.3 **a** Degree of aggregation by VPO at different concentrations of (*P*)-Ox-**7** (trifluoromethylbenzene, 40 °C). **b** Size distribution of (*P*)-Ox-**7** obtained by DLS experiments at various concentrations in trifluoromethylbenzene at 25 and 40 °C

light scattering (DLS) analysis (trifluoromethylbenzene; 4.0–8.0 mM) showed an average diameter of 4 nm at 25 °C and 40 °C (Fig. 2.3b). This is another example of two-atom-linker helicene oligomers, in which there are structure changes between a dimeric homo-double-helix and the monomeric form in solution. In contrast to other members of helicene oligomers [2–10], a small change was observed by CD analysis upon dimeric homoaggregation.

References

1. Okubo H, Yamaguchi M, Kabuto C (1998) Macrocyclic Amides Consisting of Helical Chiral 1,12-Dimethylbenzo[c]phenanthrene-5,8-dicarboxylate. J Org Chem 63:9500
2. Saito N, Terakawa R, Shigeno M, Amemiya R, Yamaguchi M (2011) Side Chain Effect on the Double Helix Formation of Ethynylhelicene Oligomers. J Org Chem 76:4841
3. Ichinose W, Shigeno M, Yamaguchi M (2012) Multiple States of Dimeric Aggregates Formed by (Amido–ethynyl)helicene Bidomain Compound and (Amido–ethynyl–amido)helicene Tridomain Compound. Chem Eur J 18:12644

4. Ichinose W, Ito J, Yamaguchi M (2012) Heteroaggregation between Isomeric Amido-ethynyl-amidohelicene Tridomain Oligomers. J Org Chem 77:10655
5. Miyagawa M, Yagi A, Shigeno M, Yamaguchi M (2014) Equilibrium crossing exhibited by an ethynylhelicene (*M*)-nonamer during random-coil-to-double-helix thermal transition in solution. Chem Commun 50:14447
6. Amemiya R, Ichinose W, Yamaguchi M (2010) Synthesis and Thermally Stable Helix-Dimer Formation of Amidohelicene Oligomers. Bull Chem Soc Jpn 83:809
7. Ichinose W, Miyagawa M, Ito J, Shigeno M, Amemiya R, Yamaguchi M (2011) Synthesis and duplex formation of the reverse amidohelicene tetramer. Tetrahedron 67:5477
8. Shigeno M, Kushida Y, Yamaguchi M (2013) Molecular Thermal Hysteresis in Helix-Dimer Formation of Sulfonamidohelicene Oligomers in Solution. Chem Eur J 19:10226
9. Shigeno M, Kushida Y, Yamaguchi M (2015) Self-catalysis in thermal hysteresis during random-coil to helix-dimer transition of the sulfonamidohelicene tetramer. Chem Commun 51:4040
10. Shigeno M, Sato M, Kushida Y, Yamaguchi M (2014) Aminomethylenehelicene Oligomers Possessing Flexible Two-Atom Linker Form a Stimuli-Responsive Double-Helix in Solution. Asian J Org Chem 3:797

Chapter 3
Fibril Film Formation of Pseudoenantiomeric Oxymethylenehelicene Oligomers at the Liquid–Solid Interface

Abstract A 1:1 mixture of pseudoenantiomeric oxymethylene helicene (*P*)-pentamer and (*M*)-hexamer formed a hetero-double-helix, which self-assembled into one-dimensional fibril films at liquid–solid interfaces. Discontinuous heterogeneous nucleation occurred, which involved the formation of 50-nm-diameter particles and subsequent fibril growth from these particles. The fibril film was formed on the solid surface, and the molecules remained dissociated in solution. The fibril film formation was affected by seeding and the solid surface materials.

Keywords Fibril film · Liquid–solid interface · Hetero-double-helix · Discontinuous heterogeneous nucleation–growth

Fibrils are formed by the self-assembly of organic molecules and, when fibril formation occurs in the solution phase, gels are formed by the incorporation of large amounts of solvent molecules [1–4]. In contrast, fibril formation at the liquid–solid interface results in thin fibril films and many two-dimensional crystallizations that form a sheet on the surface have been reported [5–10]. Accordingly, important biological events are induced by fibril films on the membrane, cell, and solid surfaces. For example, the formation of amyloid fibrils by the aggregation of a peptide or a protein is considered to be related to Alzheimer's, Parkinson's, and prion diseases [11, 12], in which membrane and cell surfaces affect one-dimensional fibril formation [13–15]. The surface adsorption of peptides and proteins is critical in relation to the biocompatibility of implantation materials and biofilm formation by microorganisms [16–18].

The process of fibril film formation at the liquid–solid interface is also an interesting subject. A heterogeneous nucleation–growth mechanism can be suggested in such a process, in which the initial slow formation of cluster aggregates (nucleation) of a certain size is followed by rapid polymerization (growth) [19–24]. Experimentally, sigmoidal kinetics with an initial lag time are obtained, and a seeding experiment, in which a solution is allowed to make contact with a fibril film, accelerates the process with the disappearance of the lag time. Fibril film formation can be accompanied by molecular structural changes on the solid surface, and the solid surface can catalyze molecular structural changes [25–30].

T. Sawato, *Synthesis of Optically Active Oxymethylenehelicene Oligomers and Self-assembly Phenomena at a Liquid–Solid Interface*, Springer Theses, https://doi.org/10.1007/978-981-15-3192-7_3

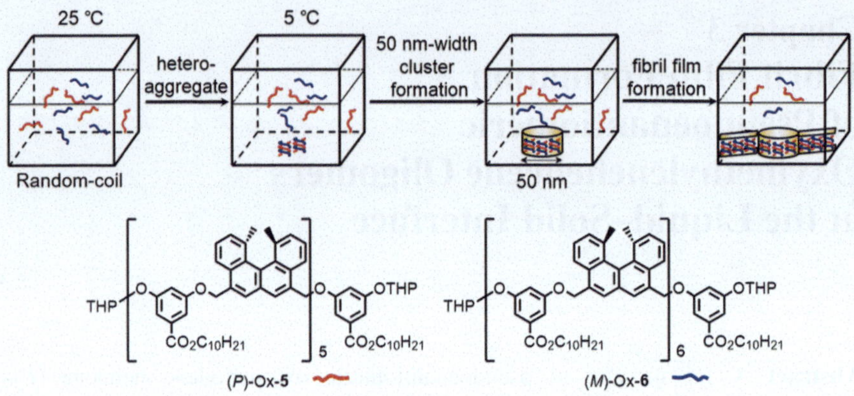

Fig. 3.1 Fibril film formation of (*P*)-Ox-**5**/(*M*)-Ox-**6** at liquid–solid interface

In this study, a pseudoenantiomeric mixture of (*P*)-Ox-**5**/(*M*)-Ox-**6** formed hetero-double-helix at the liquid–solid interface, which formed fibril films. This is a novel self-assembly of non-peptide synthetic compounds on the solid surface through discontinuous heterogeneous nucleation [31]. The hetero-double-helix self-assemble to form particles that are 50 nm in diameter at the liquid–solid interface, which then cease to grow, and fibrils start to form from the particles (Fig. 3.1). The process was examined by various methods, including spectroscopic methods, microscopic methods, seeding experiments, and surface effect experiments. Herein, I provide a systematic insight into the formation of one-dimensional fibril films at the liquid–solid interface by using synthetic compounds.

3.1 Formation of Hetero-double-helix and Fibril Film

A mixture of pseudoenantiomeric oxymethylenehelicene oligomers, being enantiomeric oligomers with different helicene numbers, formed hetero-double-helix at the liquid–solid interface, which self-assembled to form fibril films. A mixture of (*P*)-Ox-**5** and (*M*)-Ox-**6** in trifluoromethylbenzene (0.5 mM) was heated in a quartz cell at 60 °C. Very weak Cotton effects were observed (Fig. 3.2). VPO analyses (40 °C) also indicated a dissociated state: obsd: $(3.9 \pm 0.3) \times 10^3$ (total concentration, 0.5 mM), $(3.7 \pm 0.1) \times 10^3$ (total concentration, 1.0 mM); calcd: 3624. The solution was cooled to 25 °C and allowed to settle for 10 h, after which time a very small change was observed by CD analysis (Fig. 3.2). DLS analysis of the solution of (*P*)-Ox-**5**/(*M*)-Ox-**6** at 60 and 25 °C in trifluoromethylbenzene (0.5 mM) showed a small average diameter of approximately 1–2 nm for the substances in solution (Fig. 3.3).

The solution was then cooled to 5 °C and allowed to settle for 3 h, after which time strong negative and positive Cotton effects appeared at $\lambda = 316$ and 293 nm, respectively, which suggested the formation of a hetero-double-helix (Fig. 3.4). It

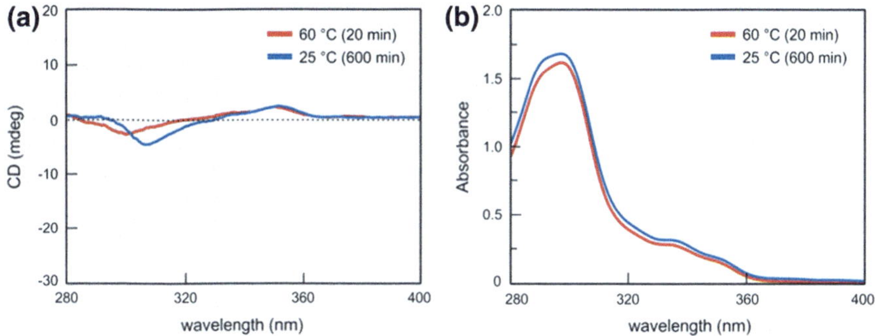

Fig. 3.2 CD (**a**) and UV-vis spectra (**b**) of (*P*)-Ox-**5**/(*M*)-Ox-**6** (1:1) in trifluoromethylbenzene (0.5 mM). A solution in a quartz cell (optical path length 0.0107 cm) was heated at 60 °C (20 min), and cooled to 25 °C (600 min)

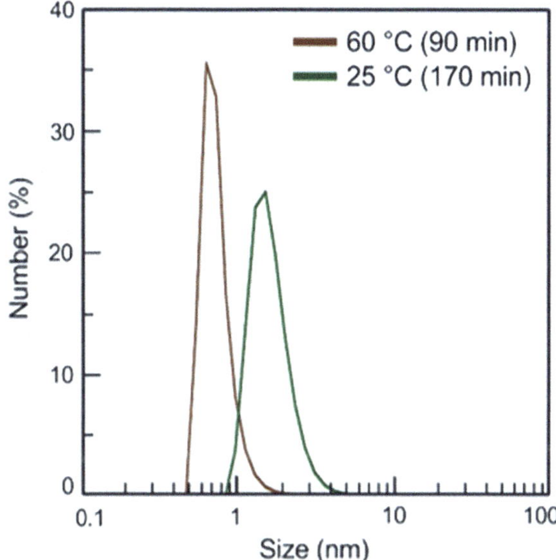

Fig. 3.3 Size distributions of (*P*)-Ox-**5**/(*M*)-Ox-**6** (1:1) obtained by DLS experiments in trifluoromethylbenzene (0.5 mM)

was also observed that thin films formed on the surface of the quartz cell. The solution phase was removed and the film and solution were separately analyzed. The yield of (*P*)-Ox-**5**/(*M*)-Ox-**6** in each phase was determined by dissolving (*P*)-Ox-**5**/(*M*)-Ox-**6** in chloroform at 25 °C and analyzing the amount of dissociated (*P*)-Ox-**5**/(*M*)-Ox-**6** by using UV/Vis calibration curves: 84% in solution and 15% on the surface. CD spectra showed no sign of hetero-double-helix in solution (Fig. 3.4a). In contrast, the CD spectra of the film on the quartz cell surface exhibited strong Cotton effects,

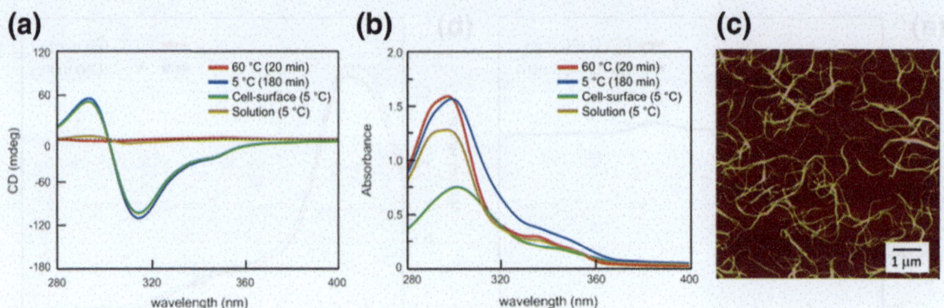

Fig. 3.4 CD (**a**) and UV/Vis (**b**) spectra of (*P*)-Ox-**5**/(*M*)-Ox-**6** in trifluoromethylbenzene (0.5 mM) for fibril film on the cell surface and solution phase. A solution (0.80 mL) in a quartz cell was heated at 60 °C for 20 min (red) and cooled to 5 °C for 180 min (blue). The solution phase was transferred into another quartz cell and immediately analyzed by CD and UV/Vis spectroscopy at 5 °C (yellow). The cell surface was analyzed by CD and UV/Vis spectroscopy at 5 °C (green). **c** AFM images (height mode) of the fibril film on a glass plate prepared in a solution of (*P*)-Ox-**5**/(*M*)-Ox-**6** in trifluoromethylbenzene (0.5 mM) at 5 °C for 60 min

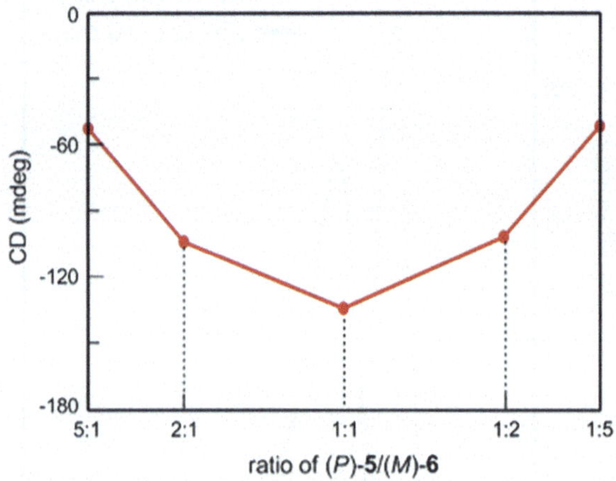

Fig. 3.5 The Job plots using CD value (316 nm) of the film on quartz cell surface against the ratio (*P*)-Ox-**5**/(*M*)-Ox-**6**. Lines were connected point to point

which were essentially identical to those in the cell containing both the solution and film (Fig. 3.4a). A Job plot experiment on the film was performed by using CD and showed 1:1 complex formation (Fig. 3.5). AFM analysis indicated the formation of fibril films with widths of 50 nm on the surface; these were not amorphous films (Fig. 3.4c). Thus, upon cooling the solution of (*P*)-Ox-**5**/(*M*)-Ox-**6** to 5 °C, a hetero-double-helix was formed on the quartz cell surface, which self-assembled to yield a fibril film. In contrast, (*P*)-Ox-**5**/(*M*)-Ox-**6** remained dissociated in solution. It was also noted that the film formed at 5 °C, but not at 25 °C. It may be reasonable to assume that hetero-double-helix of (*P*)-Ox-**5**/(*M*)-Ox-**6** is initially formed followed by self-assembly to form higher aggregates.

Fibril formation by the self-assembly of organic molecules in solution generally yields gels, and, in contrast, this study shows the formation of fibril films on solid surfaces. The difference may be due to differences between the affinity of molecules to solvents and that to the solid surface. Gels incorporate solvent molecules, whereas film formation prefers a solid surface and extrudes solvent molecules.

3.2 Discontinuous Heterogeneous Nucleation in One-Dimensional Fibril Film Formation

The present fibril film formation process is suitable for studying the self-assembly process on a solid surface because of the involvement of structural changes on the solid surface, which are accompanied by large changes in the CD spectra. An explicit experimental example of discontinuous heterogeneous nucleation on a solid surface is provided herein.

A solution of (*P*)-Ox-**5**/(*M*)-Ox-**6** in trifluoromethylbenzene (0.5 mM) was heated at 60 °C, cooled to 5 °C, and allowed to settle. The process of fibril film formation was followed by CD analysis at $\lambda = 316$ nm. The CD ($\lambda = 316$ nm)/time profile provided a sigmoidal curve with a lag time of 30 min (Fig. 3.6). The cooling temperature was changed, and sigmoidal curves were obtained at 15 and 10 °C, whereas no change was observed at 25 °C (Fig. 3.7a). The sigmoidal kinetics were observed at concentrations of 0.5 and 0.75 mM, and a small change was observed at 0.25 mM (Fig. 3.7b).

Because the above experiments show processes both on the solid surface and in solution, the phenomena were separately analyzed. After cooling for a certain period at 5 °C, the solution was removed from the cell and the fibril film and solution were analyzed by CD at $\lambda = 316$ nm (Fig. 3.4). Identical sigmoidal curves were obtained in the mixture experiment and for the fibril film in the separation experiment (Fig. 3.6).

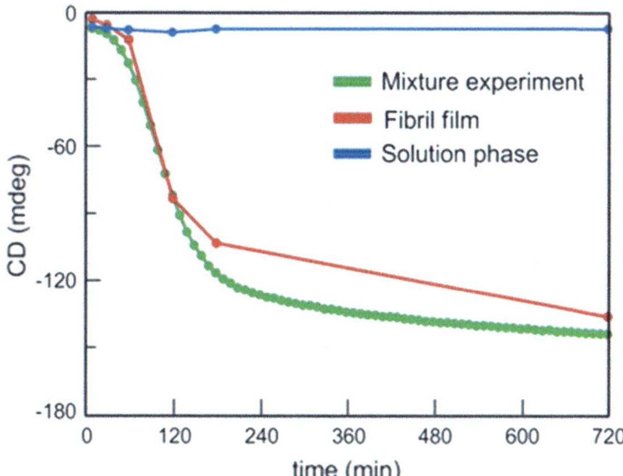

Fig. 3.6 CD (316 nm)/time profiles of (*P*)-Ox-**5**/(*M*)-Ox-**6** in trifluoromethylbenzene (0.5 mM) at 5 °C. Mixture experiment (green dots), fibril film on the cell surface (red dots), and the solution phase (blue dots) in a separation experiment. Lines are drawn between the points to guide the eye

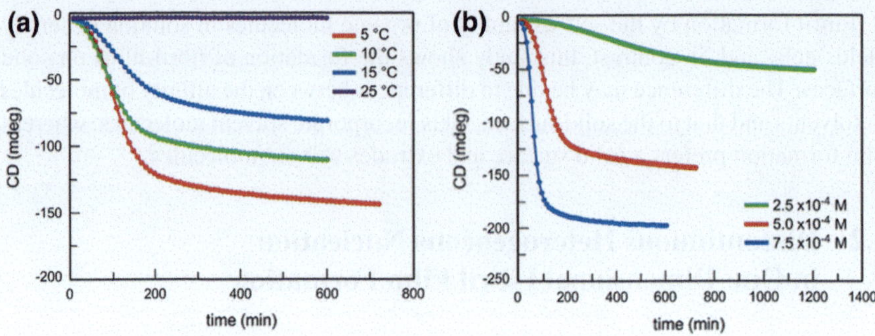

Fig. 3.7 Summary of the CD (316 nm)/time profiles of (*P*)-Ox-**5**/(*M*)-Ox-**6** in trifluoromethylbenzene. **a** Temperature effect. **b** Concentration effect. The lines are drawn between the points

In solution, no change was observed for 12 h. Quantitative UV/Vis analysis of the fibril film on the surface and in solution was also conducted, and the amount of (*P*)-Ox-**5**/(*M*)-Ox-**6** was determined by using a calibration curve (Table 3.1). After a lag time of 30 min, the yield of the fibril film increased, reached 15% after 3 h, and slowly increased to reach 29% after 12 h. Because essentially the same profiles were obtained in the mixture experiment and fibril films in the separation experiment.

AFM was employed to analyze the process of fibril film formation on the solid surface. A glass plate was dipped in a solution of (*P*)-Ox-**5**/(*M*)-Ox-**6** at 5 °C and removed after a certain period of time. After 5 min, a number of flat and round particles were observed by AFM analysis; these uniformly had a diameter of 50 nm and a height of 2–3 nm (Fig. 3.8a). It can be calculated from the volume of the particles that each particle contains approximately 250 molecules, by assuming each molecule to be 15 nm^3 [32]. The particles are critical clusters for the start of fibril formation and, notably, short fibrils form from some of the particles (Fig. 3.8b). After 10 and 15 min, many long, 50 nm wide fibrils appeared (Fig. 3.8c). Thus, fibrils grew longer without significant changes to the widths and thicknesses of the particles. This is an explicit example of discontinuous heterogeneous nucleation-growth, in which particles that are 50 nm in diameter cease to grow and start to form fibrils.

Table 3.1 Summary of the yield of (*P*)-Ox-**5**/(*M*)-Ox-**6** in fibril film and in solution phase in the separation experiment

Period of cooling at 5 °C (min)	Yield of (*P*)-Ox-**5**/(*M*)-Ox-**6** in fibril film (%)	Yield of (*P*)-Ox-**5**/(*M*)-Ox-**6** in solution (%)
10	1	98
30	2	98
60	3	95
120	9	91
180	15	84
720	29	68

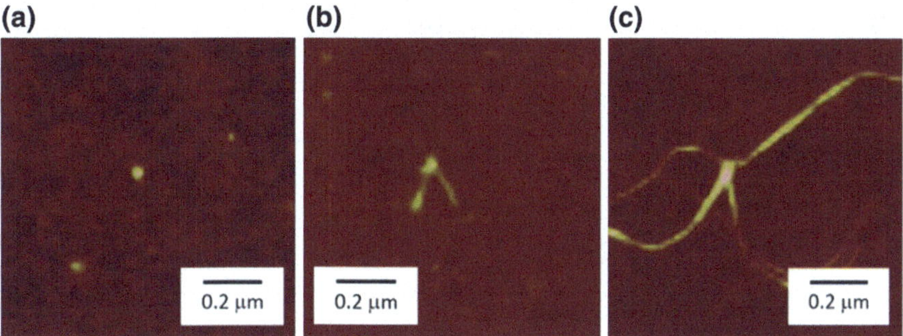

Fig. 3.8 AFM images (height mode) of the fibril film on glass-plate prepared from (*P*)-Ox-**5**/(*M*)-Ox-**6** in trifluoromethylbenzene (0.5 mM) at 5 °C

A seeding experiment showed the disappearance of the lag time for one-dimensional fibril film formation (Fig. 3.9). A solution of (*P*)-Ox-**5**/(*M*)-Ox-**6** in trifluoromethylbenzene (0.5 mM) was heated at 60 °C, cooled to 25 °C, and allowed to settle for 1 h, after which time no change was observed in the CD value (Fig. 3.10). The solution was added to another quartz cell containing a fibril film of (*P*)-Ox-**5**/(*M*)-Ox-**6**, which was prepared by allowing a solution (0.5 mM) to settle at 5 °C for 60 min and removing the solution. The CD/time profiles of the mixture at 25 °C showed an immediate decrease at 1 h, which reached −150 mdeg after 10 h. The same experiments at 40 and 50 °C also showed the disappearance of the lag time. The dissociated states of (*P*)-Ox-**5**/(*M*)-Ox-**6** in trifluoromethylbenzene (0.5 mM) at 25, 40, and 50 °C are metastable and seeding by contact with the fibril film initiates fibril film formation.

Another seeding experiment was conducted (Fig. 3.11). A fibril film was formed on a glass capillary (1 mm diameter) dipped in a solution of (*P*)-Ox-**5**/(*M*)-Ox-**6** (0.5 mM) at 5 °C for 10 h. Then, the capillary was removed from the solution and

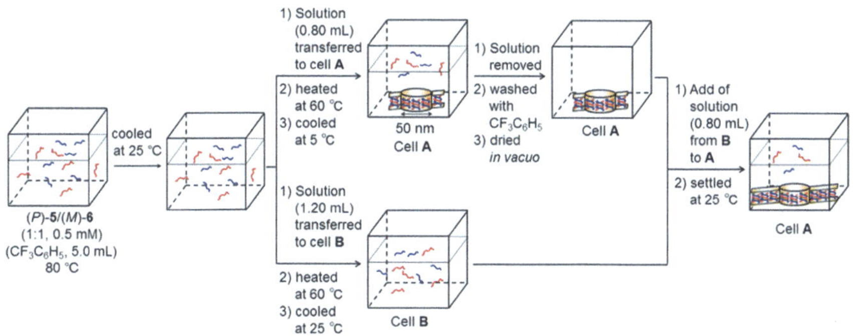

Fig. 3.9 Schematic presentation of seeding experiment at 25 °C

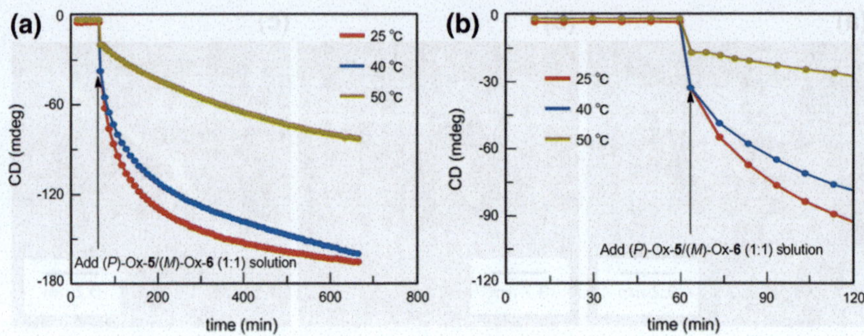

Fig. 3.10 (**a**) Seeding experiments shown by CD (316 nm)/time profiles of (*P*)-Ox-**5**/(*M*)-Ox-**6** in trifluoromethylbenzene (0.5 mM) at 25, 40, and 50 °C with fibril-film seed prepared by cooling (*P*)-Ox-**5**/(*M*)-Ox-**6** in trifluoromethylbenzene (0.5 mM) at 5 °C for 60 min. The red, blue, and yellow lines are drawn between the points. **b** Expansion of (**a**)

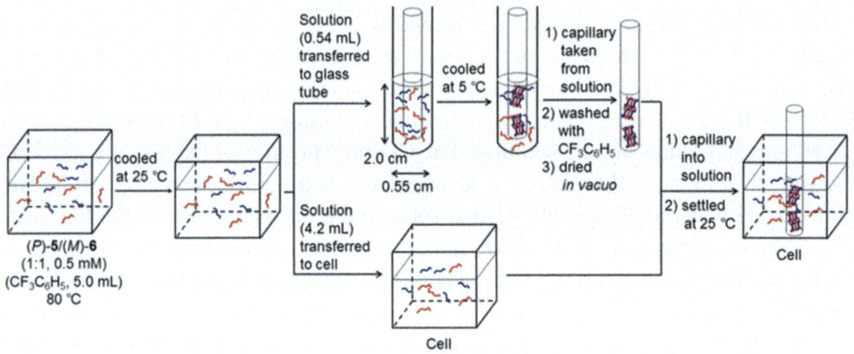

Fig. 3.11 Schematic presentation of capillary seeding experiment

allowed to settle in a quartz cell containing the dissociated solution of (*P*)-Ox-**5**/(*M*)-Ox-**6** (4.2 mL; 0.5 mM) at 25 °C, in which the capillary was not attached to the cell wall. After 24 h, essentially no film formed on the cell wall, as determined by UV/Vis analysis (Table 3.2), although the amount of fibril films on the capillary increased. The hetero-double-helix of (*P*)-Ox-**5**/(*M*)-Ox-**6** in the fibril film were not transferred

Table 3.2 Amount of (*P*)-Ox-**5**/(*M*)-Ox-**6** in the capillary seeding experiments. The initial amount of (*P*)-Ox-**5**/(*M*)-Ox-**6** on the capillary was 4–5 × 10^{-3} μmol

	Amount of (*P*)-Ox-**5**/(*M*)-Ox-**6** (μmol)		
	On the capillary surface	On the cell surface	In the solution
1st experiment	0.15	<5 × 10^{-4}	1.8
2nd experiment	0.10	<5 × 10^{-4}	1.7

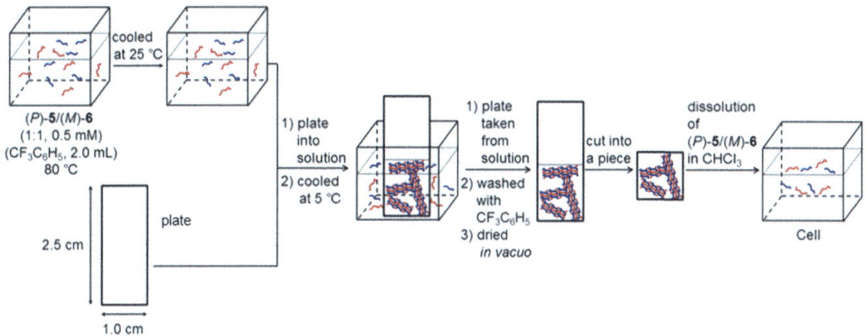

Fig. 3.12 Schematic presentation of fibril film formation on different materials of plates

from the capillary surface to the cell wall surface. Thus, nucleation of the fibril film formation does not occur in solution [33–35], but only occurs on the solid surfaces.

Fibril film formation was conducted on different material plates to determine the affinity of the hetero-double-helix (*P*)-Ox-**5**/(*M*)-Ox-**6** to the solid surfaces (Fig. 3.12). Plates with different organic and inorganic materials were put in vials containing (*P*)-Ox-**5**/(*M*)-Ox-**6** (1:1) in trifluoromethylbenzene (0.5 mM) at 5 °C for 60 min. The plates, with an area of 1 cm^2, were washed with chloroform to dissolve and dissociate (*P*)-Ox-**5**/(*M*)-Ox-**6**, and quantitative UV/Vis analysis was conducted (Table 3.3). Large amounts of fibril films formed on poly(ethylene terephthalate), polypropylene, cellulose, and gold-coated plates, and small amounts on Teflon, glass, quartz, poly(lactate), and aluminum. CD analysis of the fibril films on poly(ethylene terephthalate), gold-coated glass, glass, and quartz plates showed spectra similar to those of the fibril film in the separation experiment. The fibril film amounts differed by 20 times between poly(ethylene terephthalate) and aluminum. Fibril film formation appears to be promoted on solid surfaces with the soft nature of organic polymers

Table 3.3 Amount of (*P*)-Ox-**5**/(*M*)-Ox-**6** in the fibril films on different materials

Material	Amount of (*P*)-Ox-**5**/(*M*)-Ox-**6** in fibril film ($\times 10^{-6}$ mmol/cm^2)		
	1st experiment	2nd experiment	3rd experiment
Poly (ethylene terephthalate)	8.3	8.9	8.2
Polypropylene	4.3	3.9	3.2
Cellulose	3.6	3.4	3.8
Gold (on single-side of glass plate)	3.6	3.3	2.4
Teflon	2.3	1.5	1.3
Glass	0.9	1.4	1.3
Quartz	1.0	0.6	1.8
Poly (lactate)	0.8	0.7	0.7
Aluminium	0.3	0.4	0.3

and the gold surface. The oxymethylenehelicene oligomers (*P*)-Ox-**5**/(*M*)-Ox-**6** recognize the solid surface materials to form fibril films, which suggests the catalytic effect of the solid surfaces on aggregation and fibril formation.

A probable mechanism for the fibril film formation process by (*P*)-Ox-**5**/(*M*)-Ox-**6** is described next (Fig. 3.1). At 60 °C, (*P*)-Ox-**5**/(*M*)-Ox-**6** is in a dissociated state under equilibrium, and cooling to 25 °C produces a metastable state of dissociated (*P*)-Ox-**5**/(*M*)-Ox-**6**, with which no fibril film forms. Upon cooling to 5 °C, dimeric hetero-double-helix form on the solid surface, which self-assemble to form particles that are 50 nm in diameter. Then, the particles cease to grow and fibrils that are 50 nm start to form.

The molecular mechanism of discontinuous heterogeneous nucleation at the liquid–solid interface is discussed. A molecule of (*P*)-Ox-**5** or (*M*)-Ox-**6** is adsorbed on the solid surface, probably through van der Waals interactions. The molecule of (*P*)-Ox-**5** or (*M*)-Ox-**6** appears to favor organic surfaces or soft material surfaces, such as gold (Table 3.3). Adsorption is accompanied by a conformational change from a random coil to a folded structure, which is suitable for subsequent hetero-double-helix formation through interactions with another pseudoenantiomeric oligomer molecule. Then, the hetero-double-helix starts to self-assemble through intercomplex interactions [32, 36, 37] of the hetero-double-helix (*P*)-Ox-**5**/(*M*)-Ox-**6** with the cylindrical structure [32]. The self-assembled particles grow to 50 nm in diameter and cease; these particles probably have an ordered structure with high thermodynamic stability. It is reasonable to conclude that discontinuous nucleation is programmed into the molecular structures of (*P*)-Ox-**5**/(*M*)-Ox-**6**, and further study on the molecular mechanism is desired.

3.3 Summary

Pseudoenantiomeric mixtures of (*P*)-Ox-**5**/(*M*)-Ox-**6** formed hetero-double-helix and then fibril films formed at the liquid–solid interface. The fibril films were formed on the solid surface and not in solution. The process of fibril film formation showed sigmoidal kinetics with a lag time, and AFM analysis showed the formation of particles that were 50 nm in diameter, from which the fibrils grew. Fibril film formation was accelerated by seeding and was affected by the surface materials. This is an explicit example of fibril film formation of synthetic compounds, which occurs by discontinuous heterogeneous nucleation–growth at the liquid–solid interface.

References

1. Babu SS, Praveen VK, Ajayaghosh A (2014) Functional π-Gelators and Their Applications. Chem Rev 114:1973
2. Hirst AR, Escuder B, Miravet JF, Smith DK (2008) High-Tech Applications of Self-Assembling Supramolecular Nanostructured Gel-Phase Materials: From Regenerative Medicine to Electronic Devices. Angew Chem Int Ed 47:8002; (2008) Angew Chem 120:8122

3. Sakakibara K, Chithra P, Das B, Mori T, Akada M, Labuta J, Tsuruoka T, Maji S, Furumi S, Shrestha LK, Hill JP, Acharya S, Ariga K, Ajayaghosh A (2014) Aligned 1-D Nanorods of a π-Gelator Exhibit Molecular Orientation and Excitation Energy Transport Different from Entangled Fiber Networks. J Am Chem Soc 136:8548
4. Li T, Kalloudis M, Cardoso AZ, Adams DJ, Clegg PS (2014) Drop-Casting Hydrogels at a Liquid Interface: The Case of Hydrophobic Dipeptides. Langmuir 30:13854
5. Zhou X, Zhang Y, Zhang F, Pillai S, Liu J, Li R, Dai B, Li B, Zhang Y (2013) Hierarchical ordering of amyloid fibrils on the mica surface. Nanoscale 5:4816
6. Elliott JT, Woodward JT, Umarji A, Mei Y, Tona A (2007) The effect of surface chemistry on the formation of thin films of native fibrillar collagen. Biomaterials 28:576
7. Zhang F, Du H-N, Zhang Z-X, Ji L-N, Li H-T, Tang L, Wang H-B, Fan C-H, Xu H-J, Zhang Y, Hu J, Hu H-Y, He J-H (2006) Epitaxial Growth of Peptide Nanofilaments on Inorganic Surfaces: Effects of Interfacial Hydrophobicity/Hydrophilicity. Angew Chem In Ed 45:3611; (2006) Angew Chem 118:3693
8. Kowalewski T, Holtzman DM (1999) *In situ* atomic force microscopy study of Alzheimer's β-amyloid peptide on different substrates: New insights into mechanism of β-sheet formation. Proc Natl Acad Sci USA 96:3688
9. Lei S, Tahara K, Adisoejoso J, Balandina T, Tobe Y, de Feyter S (2010) Towards two-dimensional nanoporous networks: crystal engineering at the solid–liquid interface. CrystEngComm 12:3369
10. Ciesielski A, Palma C-A, Bonini M, Samorì P (2010) Towards Supramolecular Engineering of Functional Nanomaterials: Pre-Programming Multi-Component 2D Self-Assembly at Solid-Liquid Interfaces. Adv Mater 22:3506
11. Hamley IW (2012) The Amyloid Beta Peptide: A Chemist's Perspective. Role in Alzheimer's and Fibrillization. Chem Rev 112:5147
12. Chiti F, Dobson CM (2006) Protein Misfolding, Functional Amyloid, and Human Disease. Annu Rev Biochem 75:333
13. Cecchi C, Stefani M (2013) The amyloid-cell membrane system. The interplay between the biophysical features of oligomers/fibrils and cell membrane defines amyloid toxicity. Biophys Chem 182:30
14. Williams TL, Serpell LC (2011) Membrane and surface interactions of Alzheimer's Aβ peptide – insights into the mechanism of cytotoxicity. FEBS J 278:3905
15. Fantini J, Yahi N (2010) Molecular insights into amyloid regulation by membrane cholesterol and sphingolipids: common mechanisms in neurodegenerative diseases. Expert Rev Mol Med 12:e27
16. Schwartz K, Boles BR (2013) Microbial amyloids – functions and interactions within the host. Curr Opin Microbiol 16:93
17. Kikuchi T, Mizunoe Y, Takade A, Naito S, Yoshida S (2005) Curli Fibers Are Required for Development of Biofilm Architecture in Escherichia coli K-12 and Enhance Bacterial Adherence to Human Uroepithelial Cells. Microbiol Immunol 49:875
18. Ratner BD, Bryant SJ (2004) Biomaterials: Where We Have Been and Where We Are Going. Annu Rev Biomed Eng 6:41
19. Gillam JE, MacPhee CE (2013) Modelling amyloid fibril formation kinetics: mechanisms of nucleation and growth. J Phys: Condens Matter 25:373101
20. De Greef TFA, Smulders MMJ, Wolffs M, Schenning APHJ, Sijbesma RP, Meijer EW (2009) Supramolecular Polymerization. Chem Rev 109:5687
21. Lee C-C, Nayak A, Sethuraman A, Belfort G, McRae GJ (2007) A Three-Stage Kinetic Model of Amyloid Fibrillation. Biophys J 92:3448
22. Hovgaard MB, Dong M, Otzen DE, Besenbacher F (2007) Quartz Crystal Microbalance Studies of Multilayer Glucagon Fibrillation at the Solid-Liquid Interface. Biophys J 93:2162
23. Zhao D, Moore JS (2003) Nucleation–elongation: a mechanism for cooperative supramolecular polymerization. Org Biomol Chem 1:3471
24. Malicka JM, Sandeep A, Monti F, Bandini E, Gazzano M, Ranjith C, Praveen VK, Ajayaghosh A, Armaroli N (2013) Ultrasound Stimulated Nucleation and Growth of a Dye Assembly into Extended Gel Nanostructures. Chem Eur J 19:12991

25. Yu X, Wang Q, Lin Y, Zhao J, Zhao C, Zheng J (2012) Structure, orientation, and surface interaction of Alzheimer amyloid-β peptides on the graphite. Langmuir 28:6595
26. Yu Y-P, Zhang S, Liu Q, Li Y-M, Wang C, Besenbacher F, Dong M (2012) 2D amyloid aggregation of human islet amyloid polypeptide at the solid–liquid interface. Soft Mater 8:1616
27. Roach P, Farrar D, Perry CC (2006) Surface Tailoring for Controlled Protein Adsorption: Effect of Topography at the Nanometer Scale and Chemistry. J Am Chem Soc 128:3939
28. Lu JR, Perumal S, Hopkinson I, Webster JRP, Penfold J, Hwang W, Zhang S (2004) Interfacial Nano-structuring of Designed Peptides Regulated by Solution pH. J Am Chem Soc 126:8940
29. Knight JD, Miranker AD (2004) Phospholipid Catalysis of Diabetic Amyloid Assembly. J Mol Biol 341:1175
30. Yamamoto K, Sugiura H, Amemiya R, Aikawa H, An Z, Yamaguchi M, Mizukami M, Kurihara K (2011) Formation of double helix self-assembled monolayers of ethynylhelicene oligomer disulfides on gold surfaces. Tetrahedron 67:5972
31. Zhu M, Souillac PO, Ionescu-Zanetti C, Carter SA, Fink AL (2002) Surface-catalyzed Amyloid Fibril Formation. J Biol Chem 277:50914
32. Saito N, Shigeno M, Yamaguchi M (2012) Two-Component Fibers/Gels and Vesicles Formed from Hetero-Double-Helices of Pseudoenantiomeric Ethynylhelicene Oligomers with Branched Side Chains. Chem Eur J 18:8994
33. Mukhopadhyay RD, Ajayaghosh A (2015) Living supramolecular polymerization. Science 349:241
34. Ogi S, Stepanenko V, Sugiyasu K, Takeuchi M, Würthner F (2015) Mechanism of Self-Assembly Process and Seeded Supramolecular Polymerization of Perylene Bisimide Organogelator. J Am Chem Soc 137:3300
35. Ogi S, Sugiyasu K, Manna S, Samitsu S, Takeuchi M (2014) Living supramolecular polymerization realized through a biomimetic approach. Nat Chem 6:188
36. Yamamoto K, Oyamada N, Mizutani M, An Z, Saito N, Yamaguchi M, Kasuya M, Kurihara K (2012) Two Types of Two-Component Gels Formed from Pseudoenantiomeric Ethynylhelicene Oligomers. Langmuir 28:11939
37. Saito N, Kanie K, Matsubara M, Muramatsu A, Yamaguchi M (2015) Dynamic and Reversible Polymorphism of Self-Assembled Lyotropic Liquid Crystalline Systems Derived from Cyclic Bis(ethynylhelicene) Oligomers. J Am Chem Soc 137:6594

Chapter 4
Mechanical Stirring Induces Hetero-double-helix Formation and Self-assembly of Pseudoenantiomeric Oxymethylenehelicene Oligomers in Solution

Abstract A 1:1 mixture of pseudoenantiomeric oxymethylene helicene (*P*)-pentamers and (*M*)-hexamers formed hetero-double-helix and self-assembled as a result of mechanical stirring in solution at 25 °C. No aggregation occurred without stirring, and heteroaggregation was accelerated at high stirring rates. Repeatedly alternating between mechanical stirring and stopping respectively induced and then ceased the heteroaggregation. Sonication also promoted the heteroaggregation, which did not occur at high pressures or during centrifugation. Mechanisms based on friction due to mechanical stirring, which generated local and temporal high-temperature domains, are considered. It is shown that different methods of mechanical stimulation, namely, stirring and surface contact, induce enantiomeric heteroaggregate formation.

Keywords Mechanical stimulation · Mechanical stirring · Hetero-double-helix · Friction

Responses of organic molecules and molecular assemblies to mechanical stimulation are an interesting subject in chemistry, which has the potential for various applications such as mechanical switching, mechanical sensing, and friction control. Shearing, grinding, impact, or compression of solid organic materials to change their structures has been extensively studied with regard to the development of structural and functional materials. In biology, cells sense and respond to mechanical stimulation in solution, which controls growth, motility, and differentiation [1–3]. Thus, understanding and controlling mechanical stimulation of molecules and molecular assemblies are both critically important in various scientific fields.

Mechanical stimulation of synthetic and biological macromolecules with molecular weight larger than 5×10^3 Da in solution provided by flow, stirring, shaking, shearing, and sonication induces cleavage of covalent and noncovalent chemical bonds, which reduces molecular weight and disrupts aggregation [4–11]. In contrast, small organic molecules dispersed in solution are generally not responsive to mechanical stimulation, because hydrodynamic flow does not usually dominate the Brownian motion [4–9]. Recent studies have shown that mechanical stimulation of peptides [12–15], polyamides [16–25], surfactants [26, 27], and porphyrins [28–31]

T. Sawato, *Synthesis of Optically Active Oxymethylenehelicene Oligomers and Self-assembly Phenomena at a Liquid–Solid Interface*, Springer Theses,
https://doi.org/10.1007/978-981-15-3192-7_4

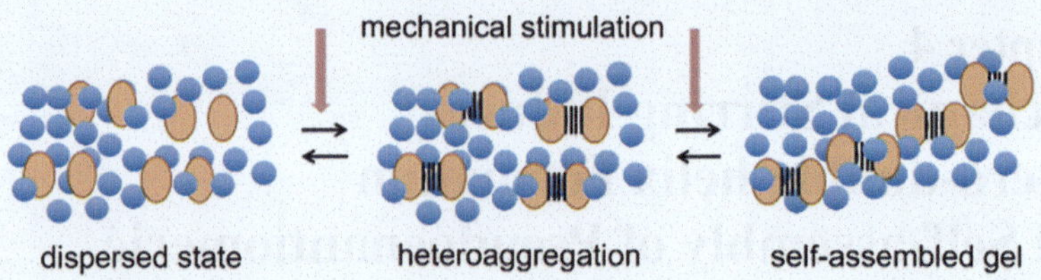

Fig. 4.1 Covalent and noncovalent chemical bond formation of molecules and subsequent self-assembly induced by mechanical stimulation in solution. Blue circles are solvent molecules, and brown circles are substrate molecules

induces the self-assembly of small molecules. Self-assembly has been ascribed in some cases to the extension and rearrangement of existing small molecular clusters.

Among such studies, an interesting responding system was developed by Otto [32], in which stirring and shaking induced the rearrangement of disulfide bonds and the self-assembly of the resulting cyclic disulfides in aqueous solution under basic conditions. The role of mechanical stimulation was discussed with regard to the breaking and growth of self-assembled fibers.

Described in this work is the hetero-double-helix of oligomeric organic compounds with molecular weights less than 5×10^3 Da that occurs as a result of mechanical stirring in an organic solvent and is followed by self-assembly (Fig. 4.1). This work provides another two-step response to mechanical stimulation, involving the formation of noncovalent chemical bonds.

It is also described here that friction between the magnetic stirring bar and the vessel surface used for mechanical stimulation is critical in the noncovalent chemical bond formation to produce hetero-double-helix. Friction is an important phenomenon in nature, which is induced by mechanical stimulation, and generates heat, electrons, and plasma; the local and temporal temperatures can reach several thousand Kelvins [4]. The resulting mechanical response originates from the generated local and temporal high-temperature domains, which causes the noncovalent chemical bonds to form, followed by self-assembly. The results suggest that magnetic stirring creates conditions in solution favorable for chemical reactions.

In the previous chapter, it was also shown that 1:1 mixtures of the pseudoenantiomeric oxymethylene helicene (*P*)-Ox-**5** and (*M*)-Ox-**6** formed hetero-double-helix at the liquid–solid interface at 5 °C which then self-assembled to form fibril films on a solid surface (Fig. 4.2). It is described in this chapter that, in response to mechanical stirring at 25 °C, hetero-double-helix were formed by the (*P*)-Ox-**5**/(*M*)-Ox-**6** system in solution, which was followed by self-assembly. It was also noted that the hetero-double-helix formation and self-assembly materials yielded enantiomeric CD spectra indicating enantiomeric structures of the hetero-double-helix. Thus, different molecular and self-assembled structures were formed in the (*P*)-Ox-**5**/(*M*)-Ox-**6** system under different conditions [33].

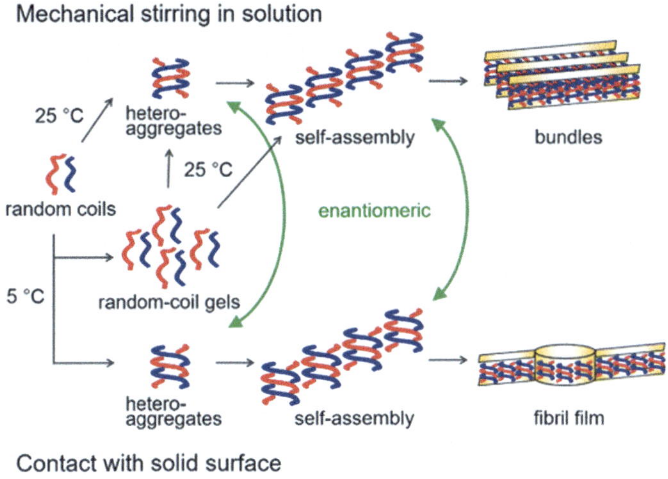

Fig. 4.2 Hetero-double-helix formation and self-assembly of a (*P*)-Ox-**5**/(*M*)-Ox-**6** mixture in solution at 25 °C by mechanical stirring compared with hetero-double-helix and fibril film formation on a solid surface at 5 °C

4.1 Stirring Induced Hetero-double-helix Formation and Self-assembly

A trifluoromethylbenzene solution (5 mL) of a mixture of (*P*)-Ox-**5** and (*M*)-Ox-**6** (0.5 mM) in a cylindrical glass vial (diameter, 21 mm) was heated at 80 °C for 3 min, and dissociated random coils were obtained [33]. Then, the solution was cooled to 25 °C to produces a metastable solution, which was mechanically stirred in a clockwise direction with an oval-shaped Teflon magnetic stirring bar (0.4 g weight) at a rate of 2000 rpm. The solution initially showed a very weak Cotton effect, which indicated the presence of the dissociated state (Fig. 4.3). The solution became turbid as the mechanical stirring was continued, and CD spectra positive for

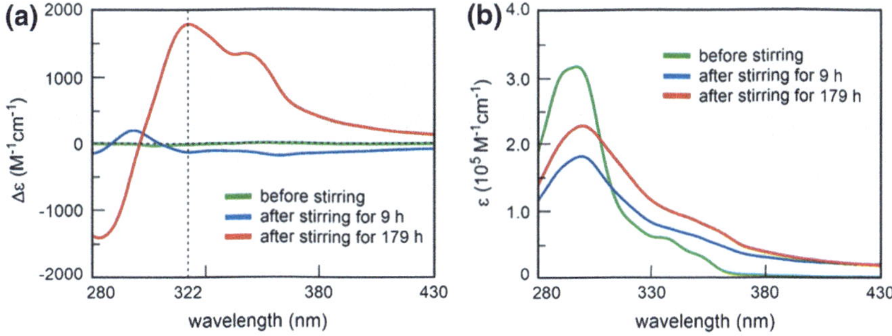

Fig. 4.3 CD (**a**) and UV-vis (**b**) spectra of (*P*)-Ox-**5**/(*M*)-Ox-**6** (0.5 mM) in trifluoromethylbenzene

the Cotton effect at 298 nm were recorded after 9 h. When the mechanical stirring was continued, the positive Cotton effect shifted to 322 nm, and the intensity increased to $\Delta\varepsilon$ + 1792 cm^{-1} M^{-1} after 179 h. DLS analysis indicated the growth of particle size (Fig. 4.4). The result indicated the hetero-double-helix formation by (*P*)-Ox-**5** and (*M*)-Ox-**6** as a result of mechanical stirring.

Formation of self-assembly materials was confirmed by AFM analysis. Immediately after cooling the solution to 25 °C, no object was observed (Fig. 4.5a). At 9 h, when $\Delta\varepsilon$ started to decrease, bundles appeared with an average width of 0.3 μm and height of 20 nm (Fig. 4.5c). These contained twisted fibers [34–36] of 50 nm diameter with a right-handed helical structure (Fig. 4.5b). The bundles formed at 116 h could be removed by filtering the solution through a 1-μm-pore membrane filter but not one with a 5-μm-pore membrane filter. A fibril film was not formed on the vessel's surface. The (*P*)-Ox-**5**/(*M*)-Ox-**6** mixture formed trifluoromethylbenzene which self-assembled to form fibers and bundles, and the solution turned turbid.

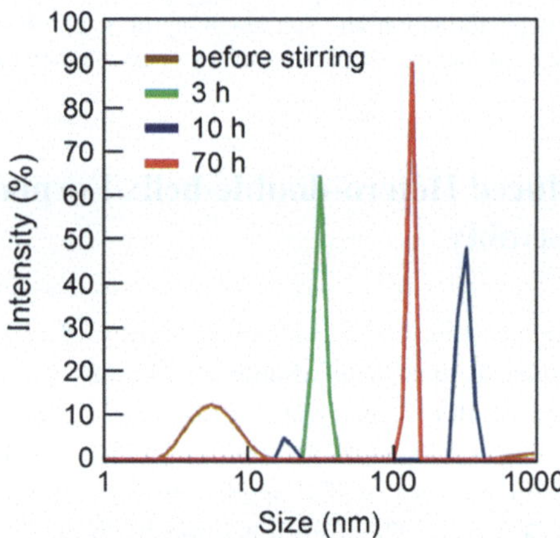

Fig. 4.4 Size distributions of (*P*)-Ox-**5**/(*M*)-Ox-**6** in trifluoromethylbenzene (0.5 mM) after stirring at the rate of 2000 rpm obtained by DLS experiments

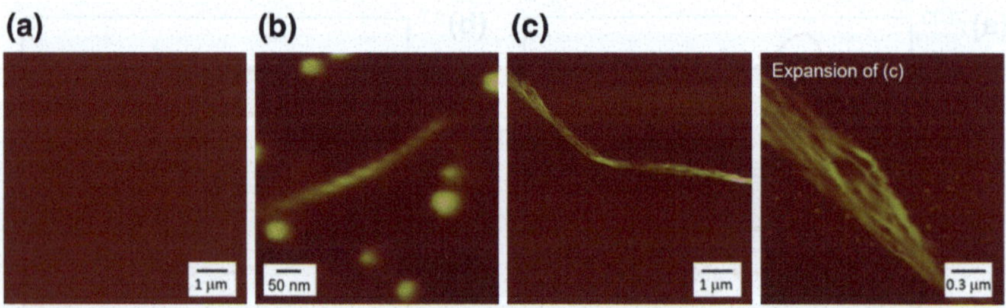

Fig. 4.5 AFM images (height mode) of (*P*)-Ox-**5**/(*M*)-Ox-**6** in trifluoromethylbenzene (0.5 mM) **a** before stirring, **b** after stirring for 5 h at 25 °C, **c** after stirring for 9 h at 25 °C

The assembly structure in the solution phase was different from that of fibril films on the solid surface that was previously reported (see Chap. 3). The metastable random-coil solution of (*P*)-Ox-**5**/(*M*)-Ox-**6** was converted to different self-assembly materials depending on the external stimulation. Notably, inverted CD spectra were obtained in these different self-assembly materials, which indicated the enantiomeric structures of the hetero-double-helix (Fig. 4.6).

The processes of heteroaggregation and self-assembly were indicated by Δε at 322 nm (Fig. 4.7). Δε remained small for 5 h, began to decrease, and reached a minimum of $-122\ \text{cm}^{-1}\ \text{M}^{-1}$ after 9 h. Then, the value started to increase and reached $+1792\ \text{cm}^{-1}\ \text{M}^{-1}$ after 179 h. This result indicated a multistep transition from random coils dispersed in solution to heteroaggregates and self-assembly materials.

Another combination of oligomers, (*P*)-heptamer and (*M*)-hexamer (*P*)-Ox-**7**/(*M*)-Ox-**6,** also provided hetero-double-helix and self-assembly materials with different CD spectra and kinetic profiles (Fig. 4.8). An advantage of this method is its applicability to various combinations of pseudoenantiomeric oligomers, which can be used to tune the hetero-double-helix formation and self-assembly properties.

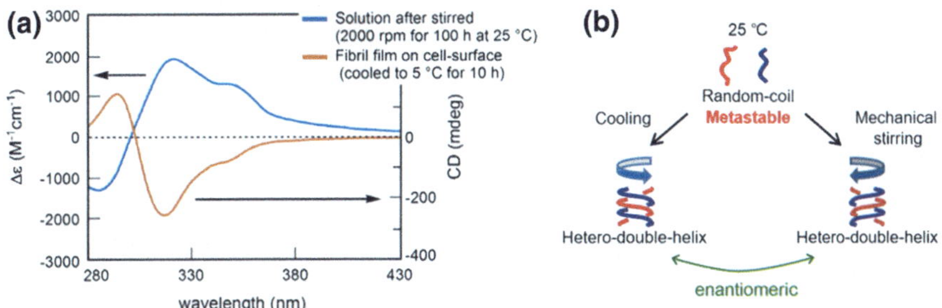

Fig. 4.6 **a** CD spectra of solution after stirring and fibril film of (*P*)-Ox-**5**/(*M*)-Ox-**6** in trifluoromethylbenzene (0.5 mM). **b** The enantiomeric molecular assembly formations by different perturbations

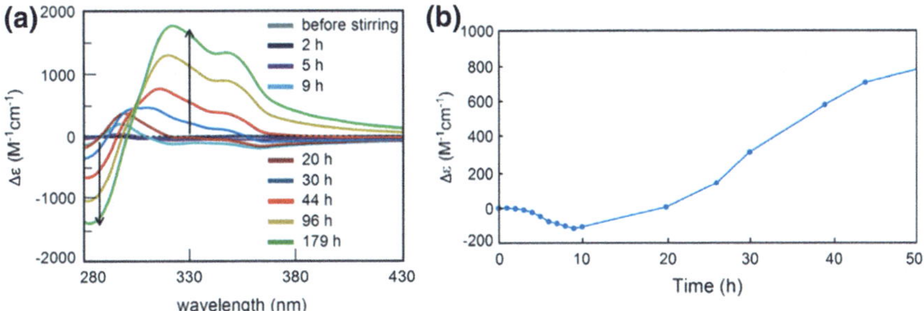

Fig. 4.7 **a** CD spectra of (*P*)-Ox-**5**/(*M*)-Ox-**6** in trifluoromethylbenzene (0.5 mM) at 25 °C. **b** Δε (322 nm)/time profiles of (*P*)-Ox-**5**/(*M*)-Ox-**6** in trifluoromethylbenzene (0.5 mM) at 25 °C

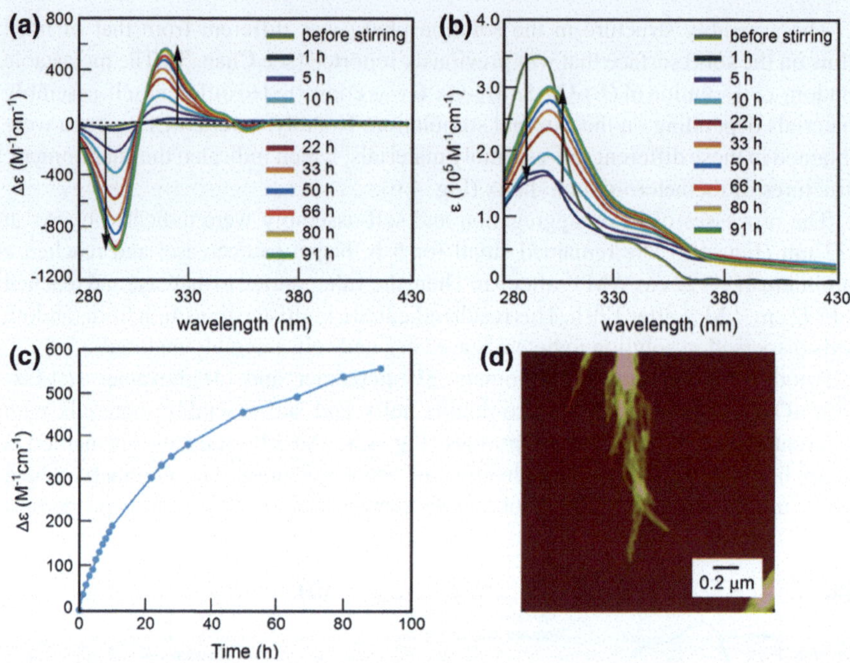

Fig. 4.8 CD (**a**) and UV–vis (**b**) spectra of (*P*)-Ox-**7**/(*M*)-Ox-**6** in a glass vial in trifluoromethylbenzene (0.5 mM) at 25 °C during stirring at 2000 rpm. **c** The Δε (322 nm)/time profiles. **d** AFM image of gels after stirring for 30 h

4.2 Effect of Mechanical Stirring Rate

The mechanical stirring rate greatly affected the formation of hetero-double-helix and self-assembly materials (Fig. 4.9). Without mechanical stirring, no aggregation occurred for the (*P*)-Ox-**5**/(*M*)-Ox-**6** system at 25 °C after 150 h. When the rate was reduced from 2000 to 1500 rpm, the process was retarded, and Δε reached + 23 cm^{-1} M^{-1} only after 40 h and +695 cm^{-1} M^{-1} after 96 h. The reversal of the stirring direction to counter-clockwise provided the same CD spectra as the clockwise experiment after 96 h (Fig. 4.10).

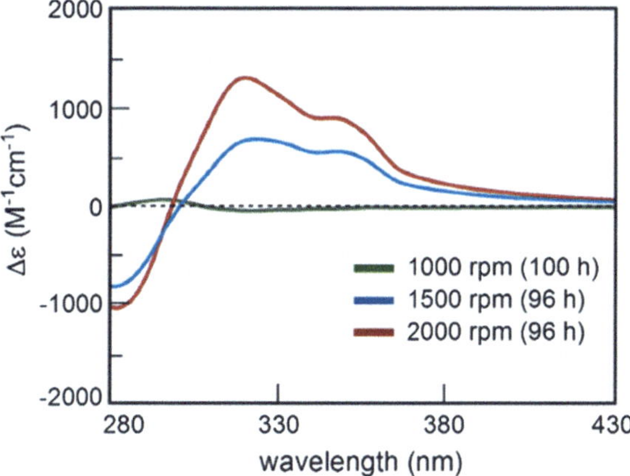

Fig. 4.9 CD spectra of (*P*)-Ox-**5**/(*M*)-Ox-**6** in trifluoromethylbenzene (0.5 mM) at 25 °C

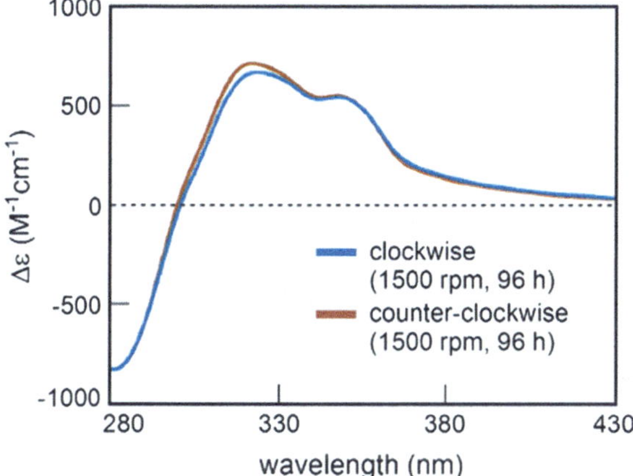

Fig. 4.10 CD spectra of (*P*)-Ox-**5**/(*M*)-Ox-**6** in trifluoromethylbenzene (0.5 mM) at 25 °C

4.3 Stop-Stirring Experiment

To probe the mechanism, mechanical stirring was stopped during the process, at which point Δε ceased to change (Fig. 4.11). The (*P*)-Ox-**5**/(*M*)-Ox-**6** solution (0.5 mM) was heated at 80 °C, cooled to 25 °C, and mechanically stirred for 10 h, during which Δε decreased to -171 cm^{-1} M^{-1}. When stirring was stopped, Δε ceased to change. At 15 h, mechanical stirring was started, and Δε began to increase, reaching $+223$ cm^{-1} M^{-1} at 35 h. Then, mechanical stirring was stopped, and Δε again ceased to change. At 40 h, mechanical stirring was restarted, which caused Δε to increase, reaching $+842$ cm^{-1} M^{-1} at 50 h. Hetero-double-helix formation occurred only during mechanical stirring.

The mechanistic aspect of the stop-stirring experiment was considered by comparing four possible cases (Fig. 4.12). In the case of mechanical stirring being a trigger or a seed inducing the formation of hetero-double-helix and self-assembly materials, the activation energy is reduced upon mechanical stimulation, and self-assembly continues to occur after mechanical stirring is stopped (Fig. 4.12a). In the case of the

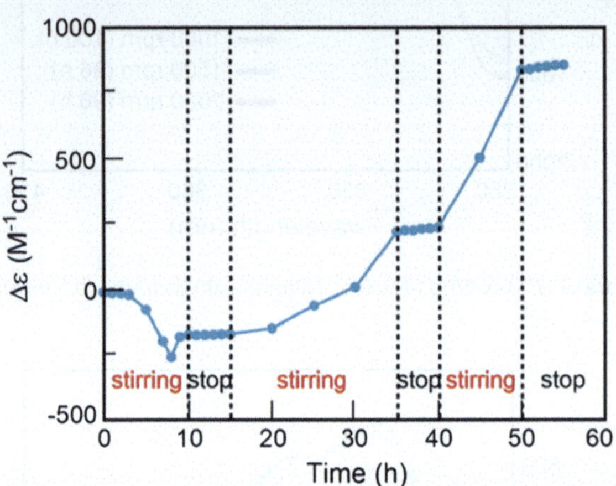

Fig. 4.11 Stop-stirring experiment with (*P*)-Ox-**5**/(*M*)-Ox-**6** in trifluoromethylbenzene (0.5 mM) at a rate of 2000 rpm

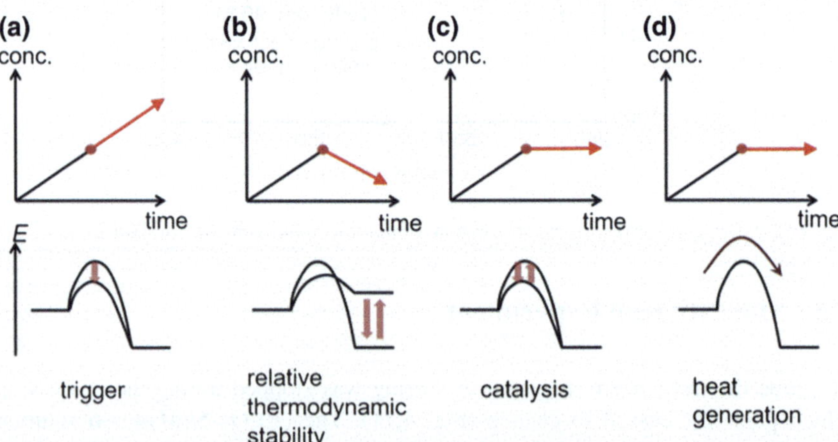

Fig. 4.12 Possible mechanisms of stop-stirring experiments. **a** Trigger mechanism: stirring triggers the decrease in activation energy, and the reaction continues after the stirring is stopped. **b** Relative thermodynamic stability mechanism: stirring changes the relative thermodynamic stability of substrates and products, and the stopping the stirring induces the formation of substrates. **c** Catalysis mechanism: stirring catalyzes the decrease in activation energy, and the reaction ceases after stopping the stirring. **d** Heat generation mechanism: stirring generates heat to overcome the activation energy barrier, and the reaction ceases after stirring is stopped

inversion of relative thermodynamic stability between the dissociated random-coil state and the hetero-double-helix/self-assembled state, stopping the stirring dissociates the hetero-double-helix (Fig. 4.12b). In the case of catalysis, activation energy is reduced as long as mechanical stirring is continued, and stopping the stirring re-establishes the high activation energy, at which point the reaction ceases (Fig. 4.12c). In the case of heat generation, the energy to overcome the activation energy barrier is provided, and stopping the stirring stops the reaction (Fig. 4.12d). The result of the stop-stirring experiment is consistent with the catalysis and heat generation mechanisms.

4.4 Effect of Vessel/Stirring Bar Materials and Shapes

The effect of the material and shape of the vessel/stirring bar was examined. The use of a cylindrical vessel (diameter, 17 mm) of tetrafluoroethylene-perfluoroalkylvinylether copolymer (PFA) and an oval Teflon magnetic stirring bar (0.4 g) slightly retarded the heteroaggregation and self-assembly, and $\Delta\varepsilon$ reached $+940\ cm^{-1}\ M^{-1}$ after 96 h (Fig. 4.13). The use of a polyethylene (diameter, 15 mm) vessel retarded the reaction, and $\Delta\varepsilon$ reached only $+336\ cm^{-1}\ M^{-1}$ after 96 h. The use of glass vessels with diameters of 12 and 21 mm provided similar results. Notably, added glass beads considerably promoted the hetero-double-helix formation and self-assembly (Fig. 4.14). Reaction in a glass vessel (diameter, 21 mm) with an oval Teflon stirring bar and glass beads (diameter, 0.5 mm; weight, 0.5 g) yielded a $\Delta\varepsilon$ value of $+1771\ cm^{-1}\ M^{-1}$ at 79 h, which was reached only after 179 h without the beads. The shape of the stirring bar was unimportant, and the use of star-shaped

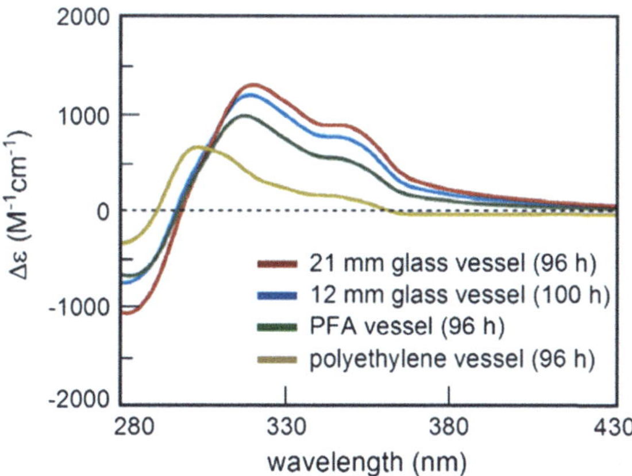

Fig. 4.13 CD spectra of (*P*)-Ox-**5**/(*M*)-Ox-**6** in trifluoromethylbenzene (0.5 mM) at 25 °C after stirring in PFA vessel (diameter, 17 mm), polyethylene vessel (diameter, 15 mm), and glass vessels (diameter, 12 and 21 mm)

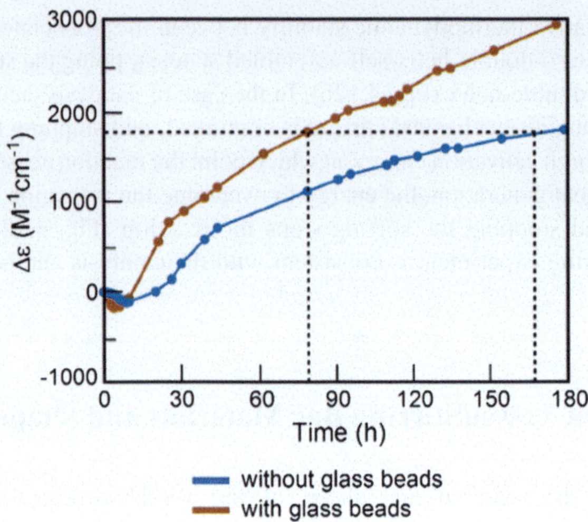

Fig. 4.14 The Δε (322 nm)/time profiles of (*P*)-Ox-**5**/(*M*)-Ox-**6** in trifluoromethylbenzene (0.5 mM) at 25 °C with and without glass beads

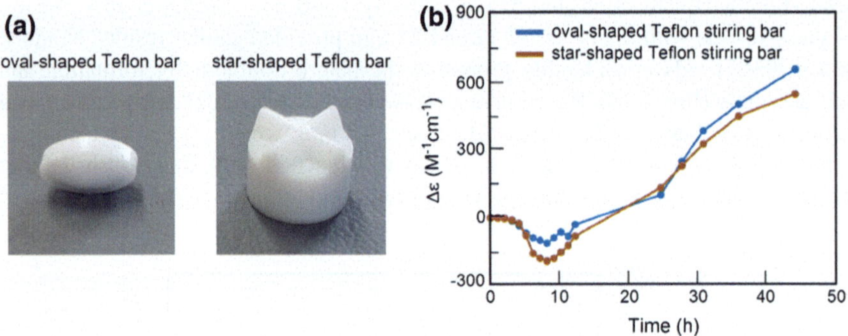

Fig. 4.15 **a** Pictures of stirring bar. **b** The Δε (322 nm)/time profiles of (*P*)-Ox-**5**/(*M*)-Ox-**6** in trifluoromethylbenzene (0.5 mM) at 25 °C using different shapes of stirring bars

Teflon bar (diameter, 10 mm; weight, 1.4 g) provided results similar to those of the oval Teflon bar (0.4 g) (Fig. 4.15). The results appear to indicate an important role of contact surface area between vessel, stirring bar, and beads.

4.5 Sonication Experiment

Related mechanical stimulations were examined. Sonication was found to be effective for the formation of hetero-double-helix and self-assembly materials (Fig. 4.16). A

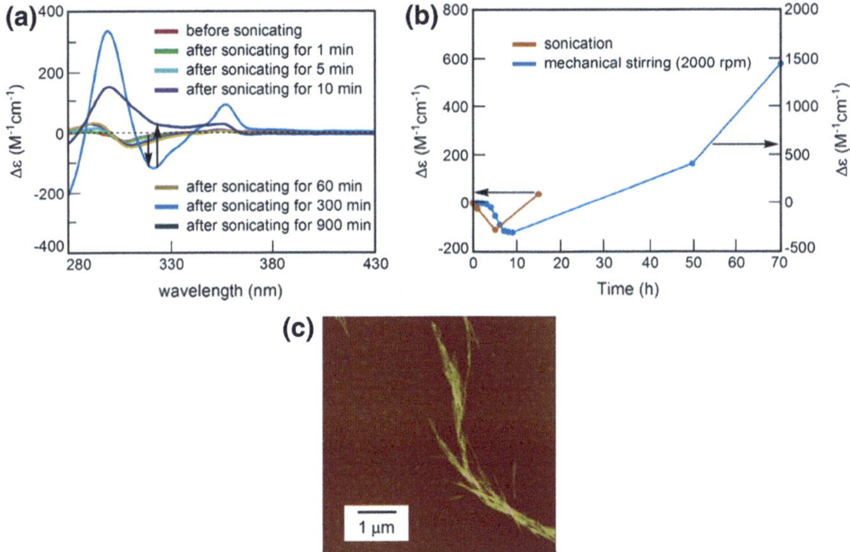

Fig. 4.16 **a** Sonication experiment of (*P*)-Ox-**5**/(*M*)-Ox-**6** (1:1) (total concentration, 0.5 mM; 25 °C; trifluoromethylbenzene) shown by CD spectra. **b** The Δε (322 nm)/time profiles of (*P*)-Ox-**5**/(*M*)-Ox-**6** in trifluoromethylbenzene (0.5 mM) at 25 °C. **c** AFM image of the dried sample after sonicating for 30 min

solution of (*P*)-Ox-**5**/(*M*)-Ox-**6** (0.5 mM) was heated at 80 °C, cooled to 25 °C, and sonicated (200 W). A small change was observed from CD spectra at 1 h, and the intensity of the peak decreased after 5 h providing a Δε (322 nm) of $-114\ cm^{-1}\ M^{-1}$. Then, the value increased to $+33\ cm^{-1}\ M^{-1}$ at 15 h, when the solution became turbid, and AFM analysis showed the formation of fibers and bundles. The process shown by the Δε (322 nm)/time profiles was similar to those obtained in the mechanical stirring experiment. When the (*P*)-Ox-**5**/(*M*)-Ox-**6** solution was subjected to a high pressure of 0.2 GPa without stirring, hetero-double-helix did not form. Centrifugation of the solution at 10,000 rpm also did not induce hetero-double-helix formation.

These results are consistent with the generation of local and temporal high-temperature domains by friction [37–40] and sonication [41, 42] (Fig. 4.17). Such heating provides sufficient energy to overcome the activation energy barrier to form hetero-double helix (Fig. 4.12). The question is then why hetero-double-helix formation occurs in response to heating, which is in contrast to the general phenomenon that molecular aggregates dissociate on heating. This result may be ascribed to the balance between the kinetic promotion of hetero-double-helix formation and the thermodynamic unfavorability of hetero-double-helix formation. The barrier between random coils and hetero-double helices is relatively high, and mechanical stirring generates local and temporal high-temperature domains, which provide the energy to form hetero-double helices (Fig. 4.12d). Rapid relaxation of the high-temperature domain occurs, and hetero-double helices diffuse in solution to form self-assembly materials.

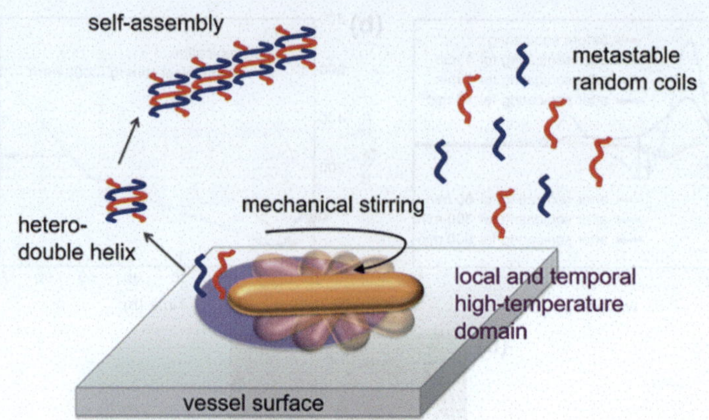

Fig. 4.17 Proposed mechanisms of hetero-double-helix formation and self-assembly of (*P*)-Ox-**5**/(*M*)-Ox-**6** by mechanical stirring

4.6 Conversion of Random-Coil Aggregates to Hetero-double-helix Aggregates by the Mechanical Stirring

The self-assembly materials can be formed via another route from the metastable random-coil (*P*)-Ox-**5**/(*M*)-Ox-**6** solution (Fig. 4.2). Previous experiment showed that the aggregation of random-coil (*P*)-Ox-**5**/(*M*)-Ox-**6** occurred without stirring at 5 °C, forming fibril films on solid surfaces (see Chap. 3). It was observed that random-coil aggregates were simultaneously formed in the solution phase (Fig. 4.18). When (*P*)-Ox-**5**/(*M*)-Ox-**6** in trifluoromethylbenzene (0.5 mM) was heated at 80 °C, cooled to 5 °C, and held at 5 °C for 10 h, fibril films were formed at the vessel surface in 36% yield, as reported previously. Aggregates were formed in the solution phase

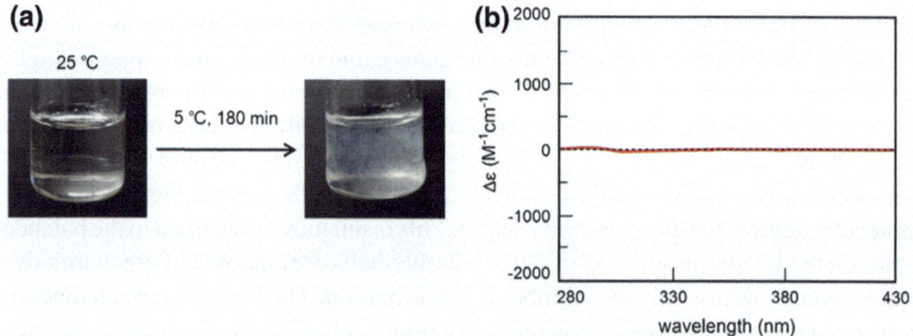

Fig. 4.18 **a** Pictures of (*P*)-Ox-**5**/(*M*)-Ox-**6** in trifluoromethylbenzene (0.5 mM) at 25 °C and after cooling at 5 °C for 180 min without stirring. **b** CD spectra of the solution phase of (*P*)-Ox-**5**/(*M*)-Ox-**6** in trifluoromethylbenzene (0.5 mM) obtained after cooling at 5 °C for 180 min

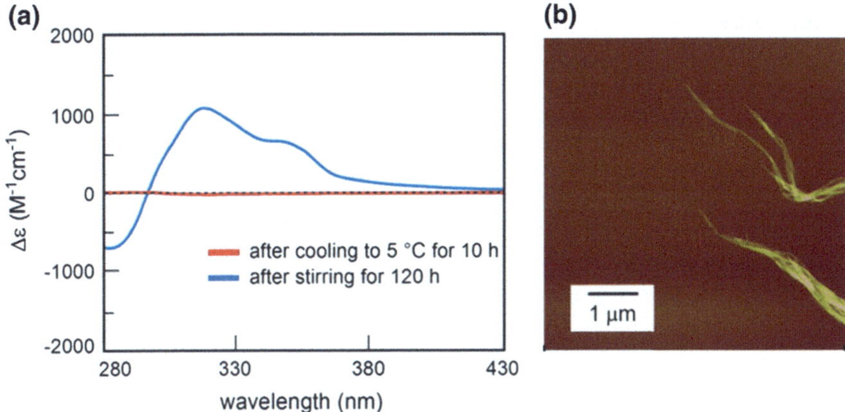

Fig. 4.19 **a** Formation of random-coil aggregates and their conversion to hetero-double-helices and self-assembly materials by mechanical stirring of (*P*)-Ox-**5**/(*M*)-Ox-**6** in trifluoromethylbenzene (0.5 mM). **b** AFM image of the dried sample after stirring for 9 h

in 41% yield as indicated by the turbid appearance and filtration experiment. CD analysis indicated the dissociated random-coil state (Fig. 4.19). The solution phase was transferred to another glass vessel (diameter, 12 mm) and mechanically stirred at 25 °C at a rate of 1500 rpm for 120 h, during which time the turbid state persisted. CD analysis indicated the formation of hetero-double-helix and self-assembly materials, similar to those in the mechanical stirring experiments. AFM analysis indicated the formation of twisted fibers 50 nm in diameter and their bundles. Thus, the mechanical stirring converted the random-coil aggregates into the self-assembly materials containing hetero-double-helix. The hetero-double-helix and self-assembly materials are in a thermodynamically stable state, which can be formed via different routes: mechanical stirring at 25 °C; settling at 5 °C, and mechanical stirring at 25 °C (Fig. 4.2).

4.7 Summary

In summary, the pseudoenantiomeric mixture of the oxymethylene helicene pentamer (*P*)-Ox-**5** and the hexamer (*M*)-Ox-**6** formed hetero-double-helix in trifluoromethylbenzene by mechanical stirring at 25 °C, which was followed by self-assembly. This is a notable example of a two-step response to mechanical stimulation involving noncovalent chemical bond formation between molecules with molecular weights below 5 $\times 10^3$ Da. The different mechanical stimulations, namely, stirring and surface contact, induce the formation of enantiomeric hetero-double-helix and self-assembly materials. Mechanisms were discussed on the basis of friction produced by mechanical stirring, which generated local and temporal high-temperature domains. Mechanical stirring provides a novel condition for reactions in solution.

References

1. Lim CT, Bershadsky A, Sheetz MP (2010) Mechanobiology. J R Soc Interface 7:S291
2. MacQueen L, Sun Y, Simmons CA (2013) Mesenchymal stem cell mechanobiology and emerging experimental platforms. J R Soc Interface 10:20130179
3. Lecuit T, Lenne P-F (2007) Cell surface mechanics and the control of cell shape, tissue patterns and morphogenesis. Nature Rev Mol Cell Biol 8:633
4. Boldyreva E (2013) Mechanochemistry of inorganic and organic systems: what is similar, what is different? Chem Soc Rev 42:7719
5. Ribas-Arino J, Marx D (2012) Covalent Mechanochemistry: Theoretical Concepts and Computational Tools with Applications to Molecular Nanomechanics. Chem Rev 112:5412
6. Beyer MK, Clausen-Schaumann H (2005) Mechanochemistry: The Mechanical Activation of Covalent Bonds. Chem Rev 105:2921
7. May PA, Moore JS (2013) Polymer mechanochemistry: techniques to generate molecular force via elongational flows. Chem Soc Rev 42:7497
8. Crusats J, El-Hachemi Z, Ribó JM (2010) Hydrodynamic effects on chiral induction. Chem Soc Rev 39:569
9. Caruso MM, Davis DA, Shen Q, Odom SA, Sottos NR, White SR, Moore JS (2009) Mechanically-Induced Chemical Changes in Polymeric Materials. Chem Rev 109:5755
10. Hara M, Komoda M, Hasei H, Yashima M, Ikeda S, Takata T, Kondo JN, Domen K (2000) A Study of Mechano-Catalysts for Overall Water Splitting. J Phys Chem B 104:780
11. Ikeda S, Takata T, Komoda M, Hara M, Kondo JN, Domen K, Tanaka A, Hosono H, Kawazoe H (1999) Mechano-catalysis—a novel method for overall water splitting. Phys Chem Chem Phys 1:4485
12. Ohhashi Y, Kihara M, Naiki H, Goto Y (2005) Ultrasonication-induced Amyloid Fibril Formation of β2-Microglobulin. J Biol Chem 280:32843
13. Fung SY, Yang H, Chen P (2007) Formation of colloidal suspension of hydrophobic compounds with an amphiphilic self-assembling peptide. Colloid Surf B: Biointerfaces 55:200
14. Helen W, de Leonardis P, Ulijn RV, Gough J, Tirelli N (2011) Mechanosensitive peptide gelation: mode of agitation controls mechanical properties and nano-scale morphology. Soft Matter 7:1732
15. Maity S, Kumar P, Haldar D (2011) Sonication-induced instant amyloid-like fibril formation and organogelation by a tripeptide. Soft Matter 7:5239
16. Shen Z, Wang T, Shi L, Tang Z, Liu M (2015) Strong circularly polarized luminescence from the supramolecular gels of an achiral gelator: tunable intensity and handedness. Chem Sci 6:4267
17. Okano K, Arteaga O, Ribo JM, Yamashita T (2011) Emergence of Chiral Environments by Effect of Flows: The Case of an Ionic Oligomer and Congo Red Dye. Chem Eur J 17:9288
18. Reddy A, Srivastava A (2014) Mechano-responsive gelation of water by a short alanine-derivative. Soft Matter 10:4863
19. Xie Z, Zhang A, Ye L, Wang X, Feng Z (2009) Shear-assisted hydrogels based on self-assembly of cyclic dipeptide derivatives. J Mater Chem 19:6100
20. van Herpt JT, Stuart MCA, Browne WR, Feringa BL (2013) Mechanically Induced Gel Formation. Langmuir 29:8763
21. Teunissen AJP, Nieuwenhuizen MML, Rodríguez-Llansola F, Palmans ARA, Meijer EW (2014) Mechanically Induced Gelation of a Kinetically Trapped Supramolecular Polymer. Macromolecules 47:8429
22. Piepenbrock M-OM, Clarke N, Steed JW (2010) Shear induced gelation in a copper(ii) metallogel: new aspects of ion-tunable rheology and gel-reformation by external chemical stimuli. Soft Matter 6:3541
23. Cravotto G, Cintas P (2009) Molecular self-assembly and patterning induced by sound waves. The case of gelation. Chem Soc Rev 38:2684
24. Liu J-W, Ma J-T, Chen C-F (2011) Structure–property relationship of a class of efficient organogelators and their multistimuli responsiveness. Tetrahedron 67:85

25. Okano K, Taguchi M, Fujiki M, Yamashita T (2011) Circularly Polarized Luminescence of Rhodamine B in a Supramolecular Chiral Medium Formed by a Vortex Flow. Angew Chem Int Ed 50:12474
26. Delgado J, Castillo R (2007) Shear-induced structures formed during thixotropic loops in dilute worm-micelle solutions. J Coll Int Sci 312:481
27. Vasudevan M, Buse E, Lu D, Krishna H, Kalyanaraman R, Shen AQ, Khomami B, Sureshkumar R (2010) Irreversible nanogel formation in surfactant solutions by microporous flow. Nat Mater 9:436
28. Tsuda A, Alam MdA, Harada T, Yamaguchi T, Ishii N, Aida T (2007) Spectroscopic visualization of vortex flows using dye-containing nanofibers. Angew Chem Int Ed 46:8198
29. Escudero C, Crusats J, Díez-Pérez I, El-Hachemi Z, Ribó JM (2006) Folding and Hydrodynamic Forces in J-Aggregates of 5-Phenyl-10,15,20-tris(4-sulfophenyl)porphyrin. Angew Chem Int Ed 45:8032
30. Ohno O, Kaizu Y, Kobayashi H (1993) *J*-aggregate formation of a water-soluble porphyrin in acidic aqueous media. J Chem Phys 99:4128
31. Ribo JM, Crusats J, Sagués F, Claret J, Rubires R (2001) Chiral Sign Induction by Vortices During the Formation of Mesophases in Stirred Solutions. Science 292:2063
32. Carnall JMA, Waudby CA, Belenguer AM, Stuart MCA, Peyralans JJ-P, Otto S (2010) Mechanosensitive self-replication driven by self-organization. Science 327:1502
33. Shigeno M, Sawato T, Yamaguchi M (2015) Fibril Film Formation of Pseudoenantiomeric Oxymethylenehelicene Oligomers at the Liquid–Solid Interface: Structural Changes, Aggregation, and Discontinuous Heterogeneous Nucleation. Chem Eur J 21:17676
34. Lin X, Kurata H, Prabhu DD, Yamauchi M, Ohba T, Yagai S (2017) Water-induced helical supramolecular polymerization and gel formation of an alkylene-tethered perylene bisimide dyad. Chem Commun 53:168
35. Yamauchi M, Ohba T, Karatsu T, Yagai S (2015) Photoreactive helical nanoaggregates exhibiting morphology transition on thermal reconstruction. Nat Commun 6:8936
36. Banno M, Yamaguchi T, Nagai K, Kaiser C, Hecht S, Yashima E (2012) Optically Active, Amphiphilic Poly(meta-phenylene ethynylene)s: Synthesis, Hydrogen-Bonding Enforced Helix Stability, and Direct AFM Observation of Their Helical Structures. J Am Chem Soc 134:8718
37. Kasem H, Brunel JF, Dufrénoy P, Siroux M, Desmet B (2011) Thermal levels and subsurface damage induced by the occurrence of hot spots during high-energy braking. Wear 270:355
38. Sutter G, Ranc N (2010) Flash temperature measurement during dry friction process at high sliding speed. Wear 268:1237
39. Rowe KG, Bennett AI, Krick BA, Sawyer WG (2013) *In situ* thermal measurements of sliding contacts. Tribol Int 62:208
40. Amiri M, Khonsari MM (2010) On the Thermodynamics of Friction and Wear—A Review. Entropy 12:1021
41. Bang JH, Suslick KS (2010) Applications of Ultrasound to the Synthesis of Nanostructured Materials. Adv Mater 22:1039
42. Xu H, Zeiger BW, Suslick KS (2013) Sonochemical synthesis of nanomaterials. Chem Soc Rev 42:2555

Chapter 5
Proximate Stochastic Chiral Symmetry Breaking by Racemic Oxymethylenehelicene Oligomers

Abstract A 50:50 mixture of oxymethylene helicene (*P*)-hexamer and (*M*)-hexamer in solution exhibited chiral symmetry breaking, which was induced by mechanical stirring, during formation of enantiomeric hetero-double-helices and their aggregates. Repeated experiments provided the negative Cotton effects, which is the deterministic chiral symmetry breaking. No change in the Cotton effect occurred in the absence of stirring. The (*P*)-hexamer to (*M*)-hexamer mixing molar fraction was varied, and a positive Cotton effect appeared at molar fractions between 40:60 and 46:54 and a negative Cotton effect at molar fractions between 48:52 and 60:40, which was reversed at 47:53. The slight deviation of symmetry from that at 50:50 was termed proximate stochastic chiral symmetry breaking. The process of chiral symmetry breaking could be tuned by varying the procedures of mechanical stirring and mixing procedures for solutions of (*P*)-hexamer and/or (*M*)-hexamer.

Keywords Chiral symmetry breaking · Racemic · Hetero-double-helix · Self-catalysis

Chiral symmetry breaking is the phenomenon in which an achiral substance is converted to a chiral substance, and an enantiomeric substance predominates [1–8]. Two processes of forming enantiomeric substances are energetically identical, and both enantiomers are generally formed in equivalent amounts [9–11]. Consider a chemical reaction to produce enantiomers **B** and ent-**B** by a bimolecular reaction of an achiral substrate 2**A**. The reaction generally provides a racemic mixture of **B** and ent-**B**. It should be noted that, when **B** or ent-**B** is predominantly formed, chiral symmetry breaking has occurred. This phenomenon is considered to involve competitive self-catalytic reactions to produce **B** and ent-**B** (Fig. 5.1a). Once the amount of one of the enantiomer **B** (or ent-**B**) is increased by chiral perturbations, the subtle difference is amplified to form a significant amount of **B** (or ent-**B**) [12–28].

It was therefore considered interesting to study the process of chiral symmetry breaking involving competitive self-catalytic reactions, especially, with regard to how the process could be affected by external perturbations during the process [29]. Repeated experiments provided information on the nature of the process. When an enantiomer **B** is much more frequently formed than ent-**B**, the phenomenon is termed deterministic chiral symmetry breaking, which may be due to persistent external

T. Sawato, *Synthesis of Optically Active Oxymethylenehelicene Oligomers and Self-assembly Phenomena at a Liquid–Solid Interface*, Springer Theses,
https://doi.org/10.1007/978-981-15-3192-7_5

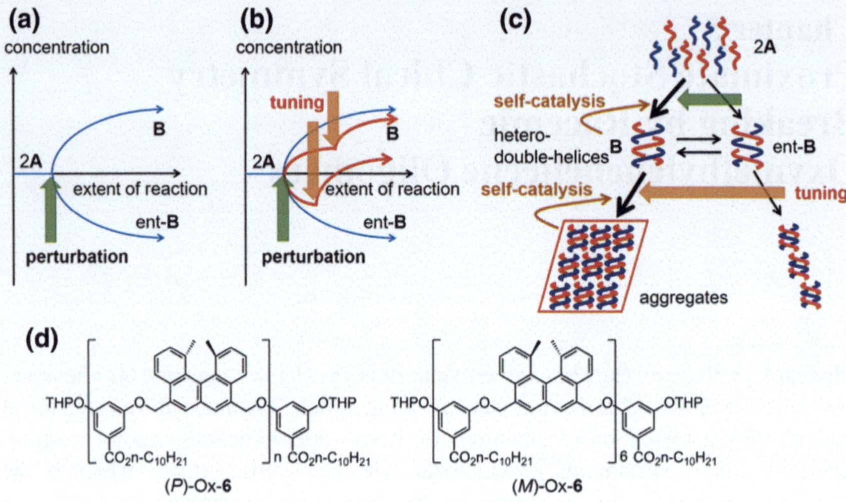

Fig. 5.1 **a** Schematic presentation of the process of chiral symmetry breaking in the reaction of random-coils 2**A** to form enantiomeric hetero-double-helices **B** or ent-**B**, expressed by a bifurcation model with external perturbation. **b** A model with mechanical tuning of the process. **c** Structural aspects of chiral symmetry breaking in this study. **d** Chemical structures of oxymethylenehelicene oligomers

perturbations. When repeated experiments provide enantiomers **B** and ent-**B** with an equal probability, the phenomenon is termed stochastic chiral symmetry breaking, which can be due to fluctuations in an environment such as temperature, pressure, or contact with chemical substances.

Described in this study is notable phenomena, which appeared during the process of chiral symmetry breaking by mechanical perturbations. The perturbations induced proximate stochastic chiral symmetry breaking, which, in this study, implies very close to stochastic chiral symmetry breaking with a slight deviation of symmetry. It was also noted that mechanical perturbations can stop and restart chiral symmetry breaking, and the enantiomeric products **B** and ent-**B** can be switched. An advantage of this system is that the initial achiral substrate is racemic, and procedures to prepare the racemic mixture can affect the process. Such study on the process of chiral symmetry breaking has not been conducted before.

A bifurcation model can be employed to describe chiral symmetry breaking involving competitive self-catalytic reactions to form **B** or ent-**B** from 2**A** (Fig. 5.1a) [2, 30–33]. Initially, a single average reaction course is taken, and, after the reaction progressed to a certain extent, an external perturbation induces a slight preference in one of the reactions to form **B** or ent-**B**. Then, the reaction is accelerated to form a significant amount of **B** or ent-**B**. When other perturbations are provided during

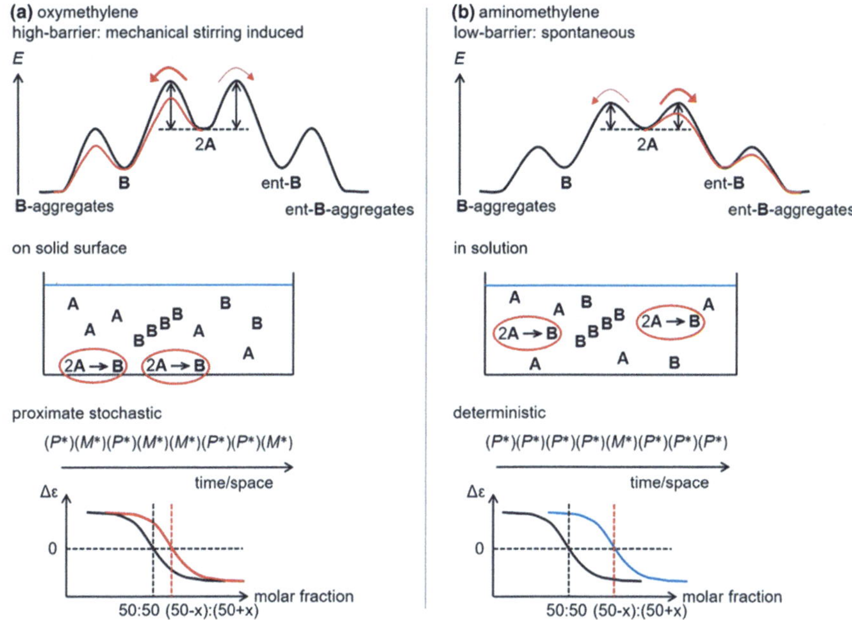

Fig. 5.2 Chiral symmetry breaking of **a** racemic oxymethylenehelicene (*P*)-Ox-**6**/(*M*)-Ox-**6** and **b** aminomethylenehelicene oligomers to form enantiomeric hetero-double-helices **B** and ent-**B** from random-coils 2**A**. The two processes are compared using energy diagrams, reaction media, and proximate stochastic/deterministic modes. In the energy diagrams, red lines indicate low barrier self-catalytic reactions. In the proximate stochastic/deterministic modes, red line indicates proximate stochastic chiral symmetry breaking; black lines indicate stochastic; blue line indicates deterministic

the competitive self-catalytic reactions, the process can further be tuned (Fig. 5.1b). This study examined the effect of mechanical tuning of chiral symmetry breaking.

We previously reported that a racemic mixture of aminomethylenehelicene (*P*)-pentamer and (*M*)-pentamer forms enantiomeric hetero-double-helices **B** and ent-**B** [34] and their self-assembled aggregates in solution, which exhibited chiral symmetry breaking (Fig. 5.1c) [29]. The competitive self-catalytic reactions are involved, in which **B** and ent-**B** catalyzes the reaction of random-coils 2**A** to become **B** and ent-**B**, respectively. Then, enantiomeric **B** and ent-**B** and their aggregates are formed [35]. Repeated experiments of the racemic (50:50) mixture provided one of the enantiomers **B** much more frequently (deterministic chiral symmetry breaking); under certain conditions, **B** and ent-**B** were formed in equal probability (stochastic chiral symmetry breaking) (Fig. 5.2b). The distance of deterministic chiral symmetry breaking from the stochastic chiral symmetry breaking was determined by repeated experiments, in which **B** and ent-**B** were formed in equal probability by non-racemic ((50 – x):(50 + x)) mixtures with significantly large x: Random selection of **B** and ent-**B** occurs by significant deviation from 50:50 mixture.

This study examined racemic mixtures of hexamers (*P*)-Ox-**6** and (*M*)-Ox-**6**, which showed proximate stochastic chiral symmetry breaking when stirred mechanically. Proximate stochastic chiral symmetry breaking is described by repeated experiments, in which both enantiomers **B** and ent-**B** form in equal probability by non-racemic ((50 – x):(50 + x)) mixtures with small x (Fig. 5.2a): Random selection of **B** and ent-**B** occurs by small deviation from 50:50 mixture. It is very close to stochastic chiral symmetry breaking, in which **B** and ent-**B** are formed in equal probability by racemic (50:50) mixtures.

It is considered that analogous competitive self-catalytic reactions are involved in the chiral symmetry breaking of the present oxymethylenehelicene oligomers (*P*)-Ox-**6**/(*M*)-Ox-**6** and aminomethylenehelicene (*P*)-pentamer/(*M*)-pentamer [29]. However, notable differences appeared in the process of chiral symmetry breaking (Fig. 5.2). The former occurred only during mechanical stirring, whereas the latter occurred spontaneously. The energy barrier must be higher in the former than in the latter, and sufficient energy to override the energy barrier is provided by mechanical stirring and friction [36]. Another difference was the reaction medium, in which chiral symmetry breaking occurred: that of (*P*)-Ox-**6**/(*M*)-Ox-**6** occurred on a solid surface, whereas that of aminomethylenehelicene oligomers occurred in bulk solution. In addition, proximate stochastic chiral symmetry breaking occurred for (*P*)-Ox-**6**/(*M*)-Ox-**6**, whereas deterministic chiral symmetry breaking occurred for aminomethylenehelicene oligomers. A notable property of (*P*)-Ox-**6**/(*M*)-Ox-**6** is that one can mechanically tune the process of chiral symmetry breaking. These mechanical tuning of chiral symmetry breaking is derived from competitive self-catalytic reactions and their high sensitivity to external perturbations [37].

5.1 Chiral Symmetry Breaking

Equal amounts of (*P*)-Ox-**6** and (*M*)-Ox-**6** were weighed using a microbalance with an accuracy of 0.5%, and dissolved in trifluoromethylbenzene (0.5 mM). The solution containing a racemic 50:50 mixture of (*P*)-Ox-**6**/(*M*)-Ox-**6** was heated to 90 °C for 10 min to prepare a solution of dissociated random-coils 2**A**, which was determined to be optically inactive by CD spectroscopy (Fig. 5.3). When the 2A solution was cooled to 25 °C, no change was observed by CD and UV-vis spectroscopy. DLS analysis (25 °C, total 0.5 mM, trifluoromethylbenzene) showed an average particle size of 2 nm. Then, the solution was mechanically stirred clockwise with a magnetic stirrer at a rate of 2000 rpm. After 10 h, the solution became turbid, and exhibited negative and positive Cotton effects at 322 and 294 nm, respectively. The intensity of the Cotton effects increased until reaching a steady state after 102 h (Fig. 5.3). The shapes of the CD spectra were similar to those of oxymethylenehelicene (*P*)-Ox-**5**/(*M*)-Ox-**6** [36, 38] and also aminomethylenehelicene (*P*)-pentamer/(*M*)-pentamer [29], which is consistent with hetero-double-helix **B** formation in racemic (*P*)-Ox-**6**/(*M*)-Ox-**6**. UV-vis spectroscopy showed a shift in the absorption maxima from 297 to 301 nm after 102 h, during which the intensity at 301 nm decreased and increased. DLS

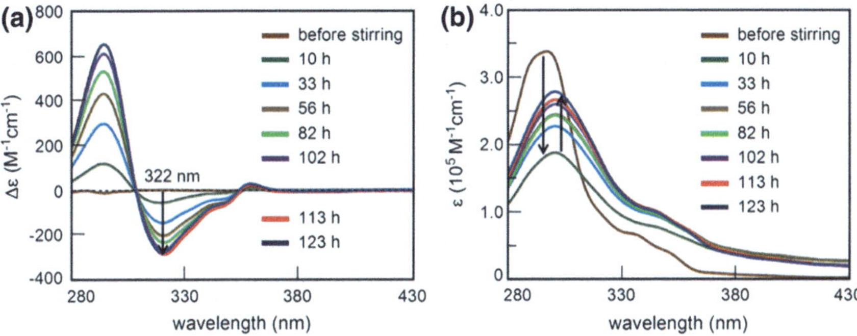

Fig. 5.3 Time courses of **a** CD and **b** UV-vis spectra in the experiments of a racemic 50:50 mixture of (*P*)-Ox-**6**/(*M*)-Ox-**6** in trifluoromethylbenzene (0.5 mM) at 25 °C

analysis (25 °C, total 0.5 mM, 40 h, trifluoromethylbenzene) of the hetero-double-helix solution showed an average particle size of 100 nm. Atomic force microscope (AFM) analysis indicated formation of aggregates (Fig. 5.4). Thus, chiral symmetry breaking appeared during the formation of **B** and their aggregation by mechanical stirring. A hetero-double-helix with a negative Cotton effect at 322 nm is referred to as **B** and its enantiomer, ent-**B**, in this study. Kinetics of chiral symmetry breaking to form **B** was examined using Δε (322 nm)/time, which yielded sigmoidal curves at the initial state (Fig. 5.5).The results is explained by the involvement of self-catalysis both in the formation of hetero-double-helices and aggregates [36].

The effect of mechanical stirring was examined for a racemic 50:50 mixture of (*P*)-Ox-**6**/(*M*)-Ox-**6** in trifluoromethylbenzene (0.5 mM). Increasing the stirring rate to 3000 rpm increased the rate of chiral symmetry breaking, and a stronger

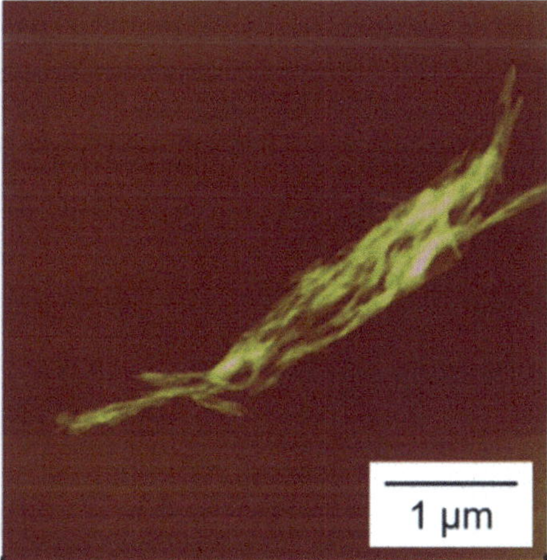

Fig. 5.4 AFM images (height mode) of a 50:50 mixture of (*P*)-Ox-**6**/(*M*)-Ox-**6** obtained from trifluoromethylbenzene solution (0.5 mM) after mechanically stirring for 100 h at 25 °C

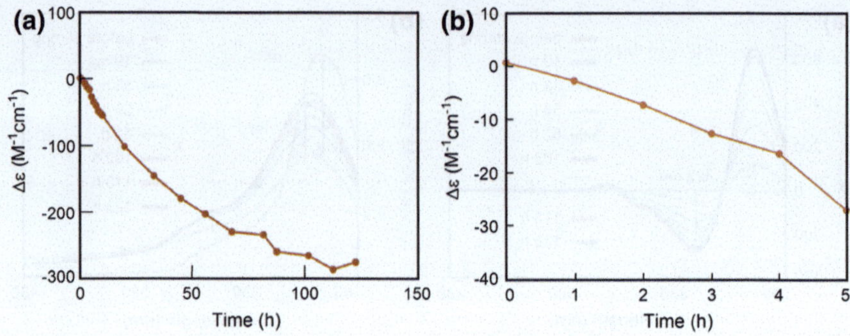

Fig. 5.5 Kinetic analysis of a 50:50 mixture of (*P*)-Ox-**6**/(*M*)-Ox-**6** in trifluoromethylbenzene (0.5 mM) at 25 °C by stirring experiment shown by Δε at 322 nm (**a**). (**b**) is expansions of (**a**). Lines are drawn to connect the points

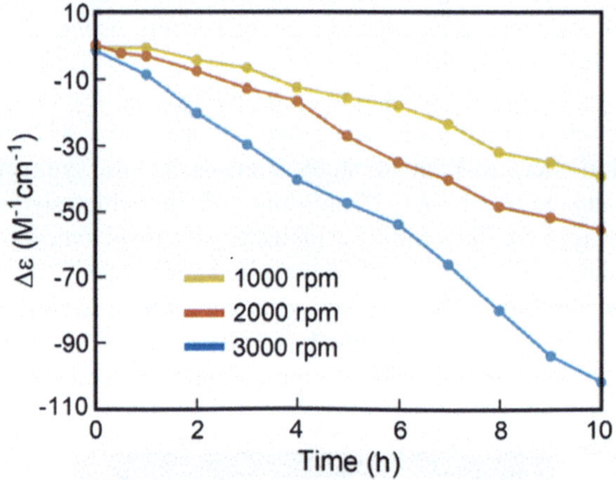

Fig. 5.6 Kinetic analysis of a 50:50 mixture of (*P*)-Ox-**6**/(*M*)-Ox-**6** in trifluoromethylbenzene (0.5 mM) at 25 °C by stirring experiments at the rate of 1000, 2000, and 3000 rpm, which are shown by Δε (322 nm)/time profiles. Lines are drawn to connect the points

negative Cotton effect appeared at 322 nm, which reached -102 cm^{-1} M^{-1} after 10 h (Fig. 5.6). Stirring in the opposite direction, namely, counterclockwise, also provided the same enantiomeric structure (Fig. 5.7). It was confirmed that no Cotton effect appeared without stirring after 265 h at 25 °C.

The material of the vessels affected the rate of chiral symmetry breaking to form **B**, and CD intensity was higher in a glass vessel than in vessels made of polyethylene or tetrafluoroethylene-perfluoroalkylvinylether copolymer (PFA) (Fig. 5.8a). Adding glass beads also accelerated the process (Fig. 5.8b). Centrifugation at 10,000 rpm did not induce chiral symmetry breaking. These observations are consistent with friction and heat generation as the driving forces for chiral symmetry breaking, as was noted previously in the oxymethylenehelicene (*P*)-**1**/(*M*)-**2** system [36].

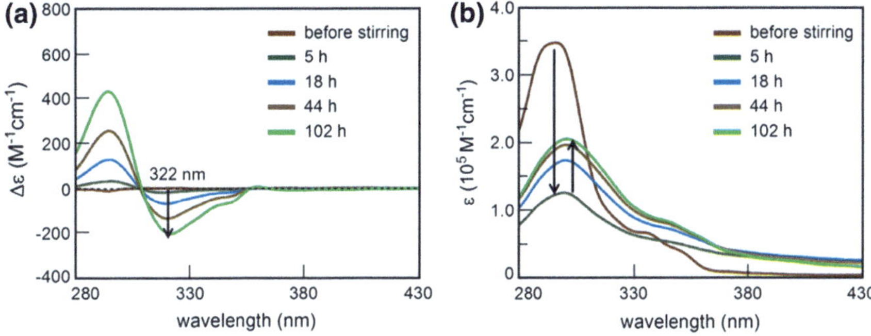

Fig. 5.7 CD (**a**) and UV-vis (**b**) spectra of a 50:50 mixture of (*P*)-Ox-**6**/(*M*)-Ox-**6** in trifluoromethylbenzene (0.5 mM) at 25 °C by counter-clockwise rotation stirring experiment

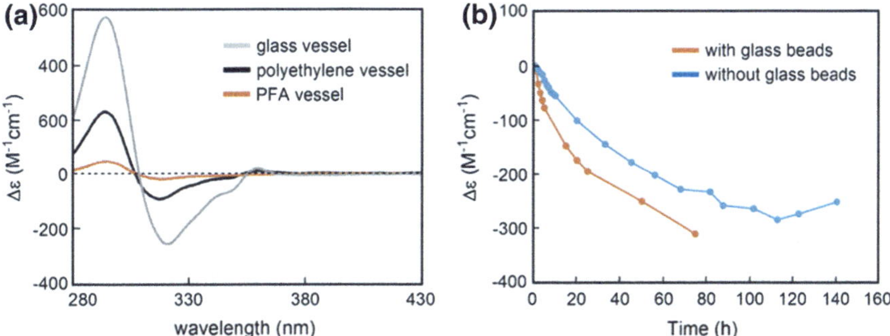

Fig. 5.8 Effects of **a** vessel materials and **b** addition of glass beads in the experiment with a racemic 50:50 mixture of (*P*)-Ox-**6**/(*M*)-Ox-**6** in trifluoromethylbenzene (0.5 mM) at 25 °C. Lines are drawn to connect the points

The effect of seeding was examined (Figs. 5.9 and 5.10). A racemic 50:50 mixture of (*P*)-Ox-**6**/(*M*)-Ox-**6** in trifluoromethylbenzene (0.5 mM) at 25 °C was stirred for 73 h to form a solution of **B**. This **B** solution (0.2 mL) was mixed with a solution of 2**A** (0.8 mL), and the resulting solution was allowed to settle for 50 min without stirring. CD spectra showed no change, and simple seeding of 2**A** with **B** did not promote the formation of **B**. Mechanical stirring was critical for the formation of **B** from 2**A**, and the process did not occur under nonstirring conditions.

To determine the stochastic and deterministic nature of chiral symmetry breaking, the experiment with racemic 50:50 mixtures was repeated (Fig. 5.11 and Table 5.1). The same lot and different lots of samples in trifluoromethylbenzene (0.5 mM) were used in 10 experiments, and all the experiments provided **B** with a negative Cotton effect at 322 nm, the Δε of which varied between −265 and −70 cm^{-1} M^{-1} (red dots in Fig. 5.11).

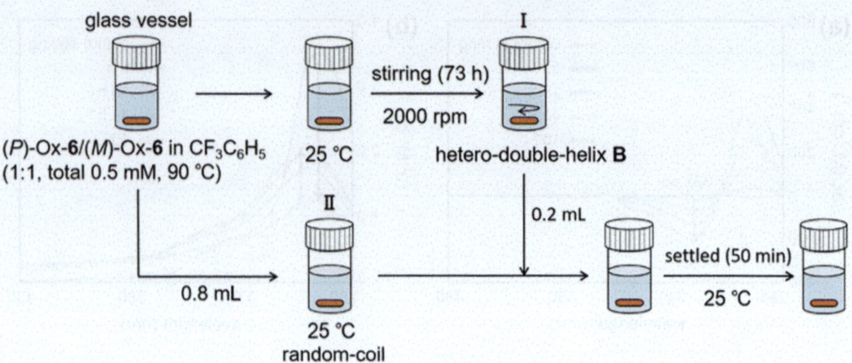

Fig. 5.9 Procedure of seeding experiment

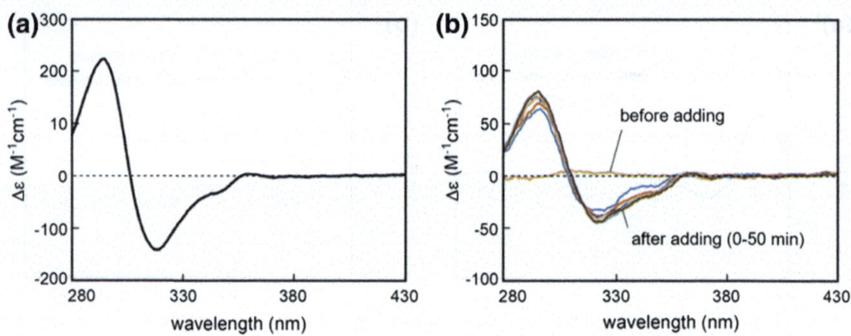

Fig. 5.10 CD spectra of a 50:50 mixture of (*P*)-Ox-**6**/(*M*)-Ox-**6** in trifluoromethylbenzene (0.5 mM) at 25 °C after stirring for 73 h (solution I) (**a**) and after mixed with random-coil solution II (**b**)

The (*P*)-Ox-**6** to (*M*)-Ox-**6** mixing molar fraction was changed (Fig. 5.11 and Table 5.2) to show proximate stochastic nature of the chiral symmetry breaking. When 49:51 and 48:52 mixtures were used in the experiments, **B** with a negative Cotton effect at 322 nm was formed predominantly in 5 experiments among 6 (orange dots). Decreasing the molar fraction to values between 46:54 and 40:60 provided the enantiomeric ent-**B** with a positive Cotton effect in all 6 experiments (green dots). The relationship between the (*P*)-Ox-**6**/(*M*)-Ox-**6** molar fraction and **B**/ent-**B** is shown by Δε/molar fraction profiles, and the experimental curve crossed Δε 0 cm^{-1} M^{-1} at approximately 47:53, which was termed proximate stochastic chiral symmetry breaking in this study (Fig. 5.2). Thus, chiral symmetry breaking of oxymethylenehelicene (*P*)-Ox-**6**/(*M*)-Ox-**6** is considerably different from that of aminomethylenehelicene oligomers [29], which showed a deterministic nature at molar fractions between 40:60 and 60:40. Although the number of experiments is not extremely large, proximate nature of chiral symmetry breaking is clearly shown, which is also indicated by following experiments.

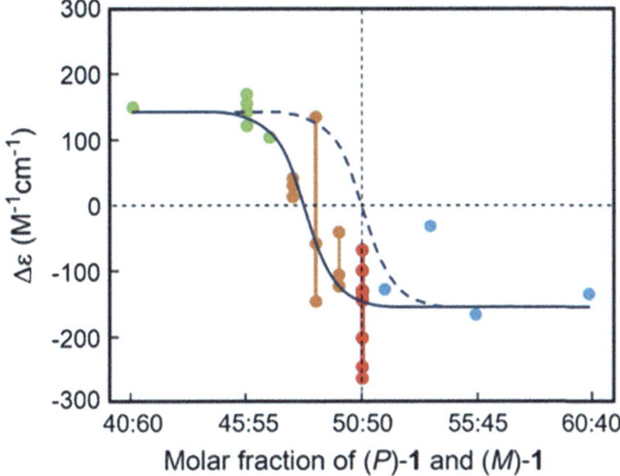

Fig. 5.11 Experiments with different molar fractions of mixtures of (*P*)-Ox-**6**/(*M*)-Ox-**6** in trifluoromethylbenzene (0.5 mM) at 25 °C, in which Δε values at 322 nm are shown. The solid line indicates experimental values showing proximate stochastic chiral symmetry breaking, and the dashed line imaginary values for stochastic chiral symmetry breaking; they are drawn to guide the eye

Table 5.1 The amounts of (*P*)-Ox-**6** and (*M*)-Ox-**6** used in repeated experiments of racemic mixtures

Lot	Exp.	Weight (mg)		Lot	Exp.	Weight (mg)	
		(*P*)-Ox-**6**	(*M*)-Ox-**6**			(*P*)-Ox-**6**	(*M*)-Ox-**6**
1	1st	0.9757	0.9776	2	2nd	0.9781	0.9765
1	2nd	0.9790	0.9787	2	3rd	0.9781	0.9766
1	3rd	0.9963	0.9777	3	1st	0.9792	0.9786
1	4th	0.9795	0.9784	3	2nd	0.9765	0.9801
2	1st	0.9771	0.9771	3	3rd	0.9790	0.9796

10 samples were prepared from three different lots of samples. (*P*)-Ox-**6** and (*M*)-Ox-**6** were weighed in a vial, and were dissolved in trifluoromethylbenzene (1.0 mL, 0.5 mM)

The nature of chiral symmetry breaking was further examined by experiments to form racemic 50:50 mixtures of (*P*)-Ox-**6**/(*M*)-Ox-**6** using different mixing procedures (Figs. 5.12, 5.13 and 5.14). A 55:45 mixture of (*P*)-Ox-**6**/(*M*)-Ox-**6** in trifluoromethylbenzene (0.5 mL, 0.5 mM) was mechanically stirred for 50 h, which produced **B** with a strong negative Cotton effect at 322 nm (Figs. 5.12 and 5.14a). The solution of **B** was mixed with a solution of random-coil 2**A** (0.5 mL, total 0.5 mM) containing a 45:55 mixture of (*P*)-Ox-**6**/(*M*)-Ox-**6**. The resulting racemic 50:50 mixture of (*P*)-Ox-**6**/(*M*)-Ox-**6** initially showed a weak negative Cotton effect at 322 nm, and mechanical stirring for 45 h resulted in a strong negative Cotton effect because of the formation of **B**. An enantiomeric reaction was conducted, in which an ent-**B** solution (0.5 mL, 0.5 mM) containing a 45:55 mixture of (*P*)-Ox-**6**/(*M*)-Ox-**6** was

Table 5.2 The amounts of (*P*)-Ox-**6** and (*M*)-Ox-**6** used in the experiments with different molar fraction of (*P*)-Ox-**6**/(*M*)-Ox-**6** mixtures

(*P*)-Ox-**6**/(*M*)-Ox-**6**	Exp.	Weight (mg)		(*P*)-Ox-6/(M)-Ox-6	Exp.	Weight (mg)	
		(*P*)-Ox-**6**	(*M*)-Ox-**6**			(*P*)-Ox-**6**	(*M*)-Ox-**6**
40:60	1st	0.7846	1.1773	48:52	2nd	0.9359	1.0196
45:55	1st	0.8816	1.0777	48:52	3rd	0.9388	1.0139
45:55	2nd	0.8793	1.0767	49:51	1st	0.9574	0.9981
45:55	3rd	0.8795	1.0754	49:51	2nd	0.9606	0.9960
45:55	4th	0.8818	1.0753	49:51	3rd	0.9569	0.9970
46:54	1st	0.8991	1.0532	51:49	1st	0.9993	0.9569
47:53	1st	0.9211	1.0353	53:47	1st	1.0403	0.9197
47:53	2nd	0.9180	1.0354	55:45	1st	1.0726	0.8791
47:53	3rd	0.9183	1.0362	60:40	1st	1.1758	0.7812
48:52	1st	0.9383	1.0132				

(*P*)-Ox-**6** and (*M*)-Ox-**6** were weighed in a vial, and were dissolved in trifluoromethylbenzene (1.0 mL, 0.5 mM)

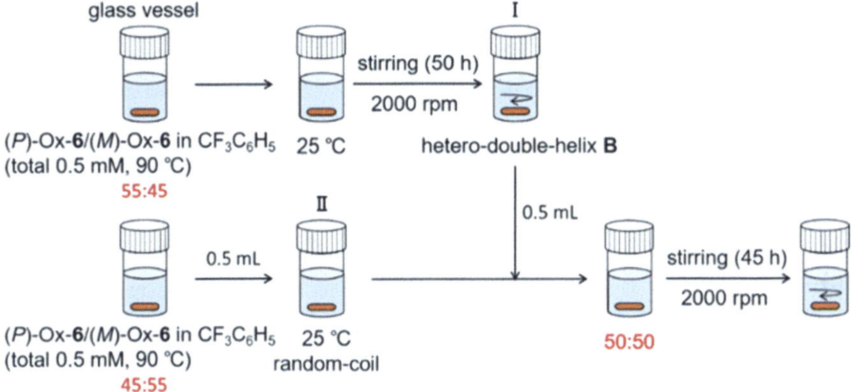

Fig. 5.12 Procedure of mixing experiment of hetero-double-helix **B** (55:45) and random-coil (45:55)

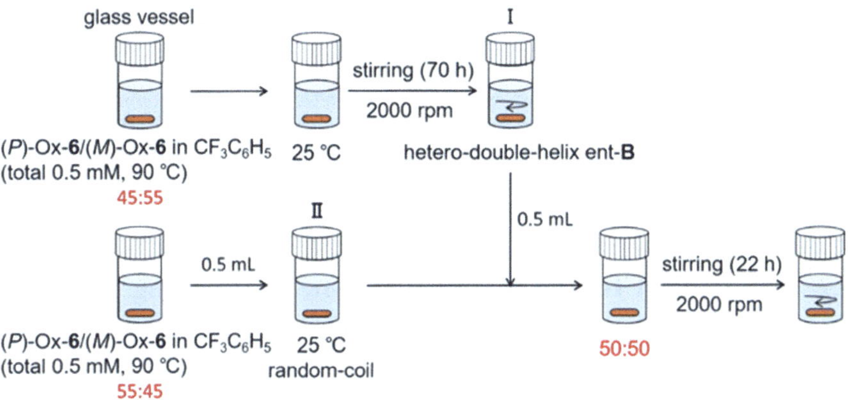

Fig. 5.13 Procedure of mixing experiment of hetero-double-helix ent-**B** (45:55) and random-coil (55:45)

formed by stirring for 70 h (Figs. 5.13 and 5.14c). The solution of ent-**B** was mixed with a solution of random-coil 2**A** (0.5 mL, 0.5 mM) containing a 55:45 mixture of (*P*)-Ox-**6**/(*M*)-Ox-**6**. The resulting racemic 50:50 mixture of (*P*)-Ox-**6**/(*M*)-Ox-**6** initially showed a weak positive Cotton effect at 322 nm, and mechanical stirring for 22 h resulted in a strong positive Cotton effect appeared, because of the formation of ent-**B**. The mixing experiments are symmetric with regard to chirality, and showed stochastic chiral symmetry breaking: see CD spectra in Fig. 5.14a, c.

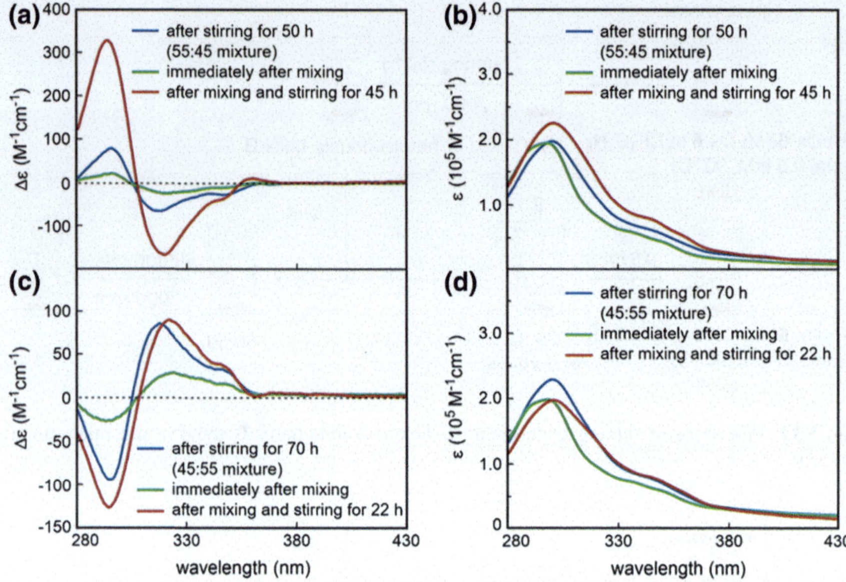

Fig. 5.14 **a** CD spectra and **b** UV-vis spectra of (*P*)-Ox-**6**/(*M*)-Ox-**6** in trifluoromethylbenzene (0.5 mM) at 25 °C formed in the mixing experiment with hetero-double-helix **B** (55:45) and random-coil 2**A** (45:55). **c** CD spectra and **d** UV-vis spectra of (*P*)-Ox-**6**/(*M*)-Ox-**6** in trifluoromethylbenzene (0.5 mM) at 25 °C formed by mixing experiment of hetero-double-helix ent-**B** (45:55) and random-coil 2**A** (55:45). CD spectra (red lines) in (**a**) and (**c**) are symmetric

5.2 Mechanical Tuning of Chiral Symmetry Breaking Process

The properties of (*P*)-Ox-**6**/(*M*)-Ox-**6** can be utilized to mechanically tune the process of chiral symmetry breaking. The effect of mechanical stirring was employed to stop and restart the process, as shown by Δε/time profiles (Fig. 5.15). When a racemic 50:50 mixture of (*P*)-Ox-**6**/(*M*)-Ox-**6** in trifluoromethylbenzene (0.5 mM) was mechanically stirred at 2000 rpm for 5 h, Δε at 322 nm decreased, reaching $-28\ cm^{-1}M^{-1}$. Stirring was then stopped, and Δε ceased to decrease. Mechanical stirring at 2000 rpm was then started, which caused Δε to decrease with approximately the same slope and to reach $-58\ cm^{-1}M^{-1}$ after 15 h. The same experiments were repeated, and chiral symmetry breaking occurred only during mechanical stirring. When mechanical stirring was stopped, the reaction stopped, which could be restarted by resuming mechanical stirring.

The mixing experiments to form racemic 50:50 mixtures in the previous section added (*P*)-Ox-**6**/(*M*)-Ox-**6** mixtures to form racemic 50:50 mixtures of (*P*)-Ox-**6**/(*M*)-Ox-**6**, which exhibited stochastic chiral symmetry breaking (Fig. 5.14). It was noted that the addition of (*P*)-Ox-**6** or (*M*)-Ox-**6** alone in place of (*P*)-Ox-**6**/(*M*)-Ox-**6** mixtures exhibited proximate stochastic chiral symmetry breaking (Figs. 5.16

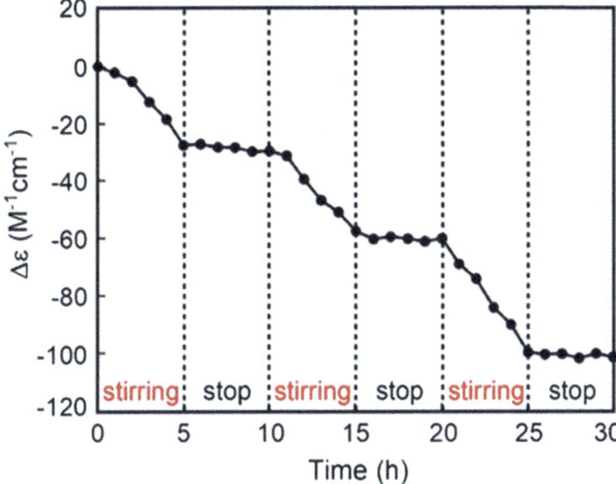

Fig. 5.15 Stop-stirring experiment of a racemic 50:50 mixture of (*P*)-Ox-**6**/(*M*)-Ox-**6** in trifluoromethylbenzene (0.5 mM) at 25 °C shown by Δε values at 322 nm. Lines are drawn to connect the points

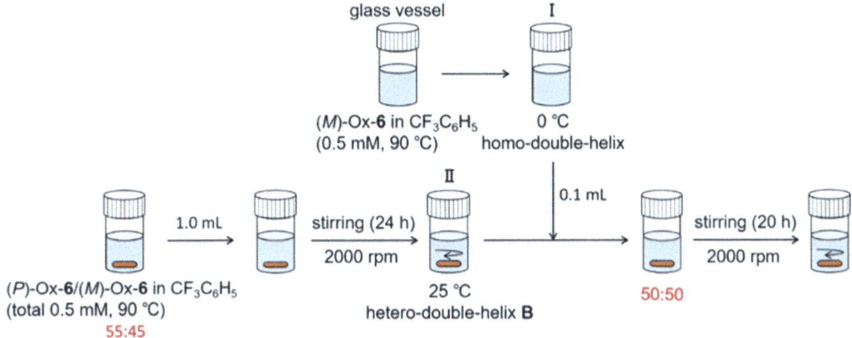

Fig. 5.16 Procedure of mixing experiment of hetero-double-helix **B** (55:45) and homo-double-helix (*M*)-Ox-**6**

and 5.17). Mechanical stirring of a 55:45 mixture of (*P*)-Ox-**6**/(*M*)-Ox-**6** in trifluoromethylbenzene (1.0 mL, 0.5 mM) at 25 °C for 24 h provided a solution of **B** with a negative Cotton effect at 322 nm (Fig. 5.17a). Homo-double-helix (*M*)-Ox-**6** solution (0.1 mL, total 0.5 mM) was then mixed to prepare a racemic 50:50 mixture of (*P*)-Ox-**6**/(*M*)-Ox-**6**. The solution was mechanically stirred at 25 °C for 20 h, and a negative Cotton effect reappeared owing to the formation of **B**.

Notably, the enantiomeric system exhibited an unsymmetric phenomenon. Mechanical stirring of a 45:55 mixture of (*P*)-Ox-**6**/(*M*)-Ox-**6** in trifluoromethylbenzene (1.0 mL, 0.5 mM) at 25 °C for 25 h provided an ent-**B** solution with a positive Cotton effect at 322 nm (Figs. 5.18 and 5.19a). A homo-double-helix (*P*)-Ox-**6** solution (0.1 mL, 0.5 mM) was then mixed to form a racemic 50:50 mixture

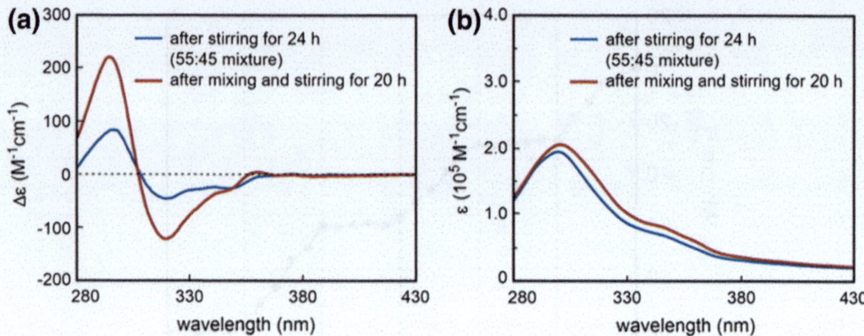

Fig. 5.17 CD (**a**) and UV-vis (**b**) spectra of (*P*)-Ox-**6**/(*M*)-Ox-**6** in trifluoromethylbenzene (0.5 mM) at 25 °C formed by the mixing experiment of hetero-double-helix **B** (55:45) and homo-double-helix (*M*)-Ox-**6**

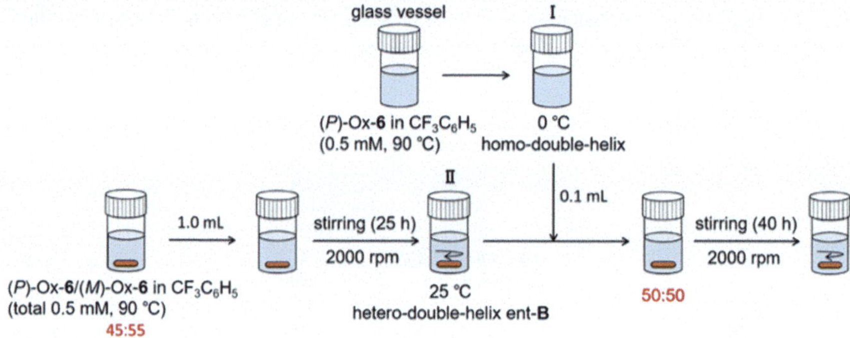

Fig. 5.18 Procedure of mixing experiment of hetero-double-helix ent-**B** (45:55) and homo-double-helix (*P*)-Ox-**6**

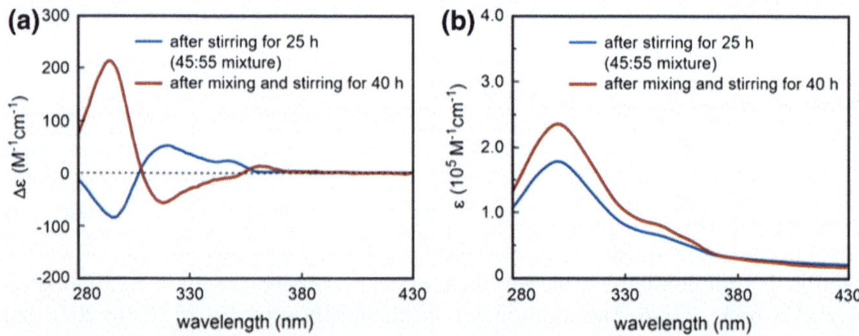

Fig. 5.19 CD (a) and UV-vis (b) spectra of (*P*)-Ox-**6**/(*M*)-Ox-**6** in trifluoromethylbenzene (0.5 mM) at 25 °C formed by the mixing experiment of hetero-double-helix ent-**B** (45:55) and homo-double-helix (*P*)-Ox-**6**

of (*P*)-Ox-**6**/(*M*)-Ox-**6**. The solution was mechanically stirred at 25 °C for 40 h, and the Cotton effect was reversed to be negative owing to the formation of **B**. Thus, the process of chiral symmetry breaking was switched between the enantiomeric hetero-double-helix products ent-**B** and **B**, which reversed the helical sense of the hetero-double-helix [34].

The above experiments employed a solution of homo-double-helix (*P*)-Ox-**6** or (*M*)-Ox-**6** to form racemic 50:50 mixtures of (*P*)-Ox-**6**/(*M*)-Ox-**6**. Then, a solution of random-coil (*P*)-Ox-**6** or (*M*)-Ox-**6** was employed, which was prepared by heating at 60 °C. The resulting racemic 50:50 mixture of (*P*)-Ox-**6**/(*M*)-Ox-**6** was mechanically stirred at 25 °C, and **B** was formed either using random-coil (*P*)-Ox-**6** or (*M*)-Ox-**6** (Figs. 5.20, 5.21, 5.22 and 5.23). Then, the same unsymmetric phenomenon occurred by adding (*P*)-Ox-**6** or (*M*)-Ox-**6** in homo-double-helix or random-coil form to (*P*)-Ox-**6**/(*M*)-Ox-**6** mixtures. The proximate stochastic nature of chiral symmetry breaking appeared. It was also noted that procedures adding

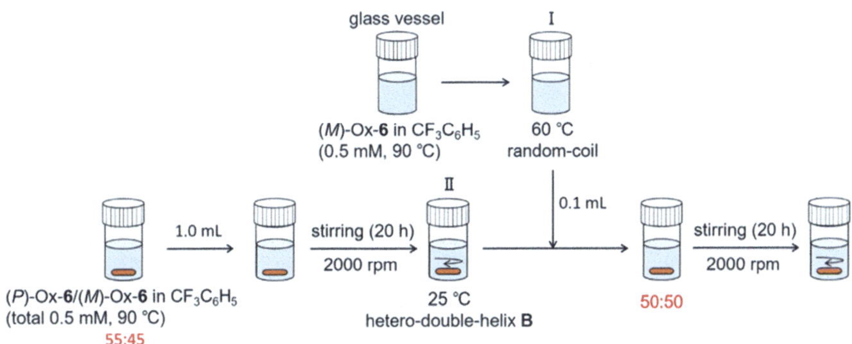

Fig. 5.20 Procedure of mixing experiment of hetero-double-helix **B** (55:45) and random-coil (*M*)-Ox-**6**

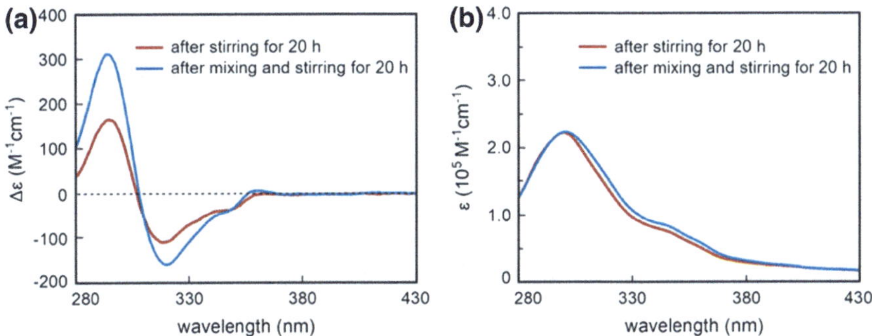

Fig. 5.21 CD spectra (**a**) and UV-vis spectra (**b**) of (*P*)-Ox-**6**/(*M*)-Ox-**6** in trifluoromethylbenzene (0.5 mM) at 25 °C formed by mixing experiment of hetero-double-helix **B** (55:45) and random-coil (*M*)-Ox-**6**

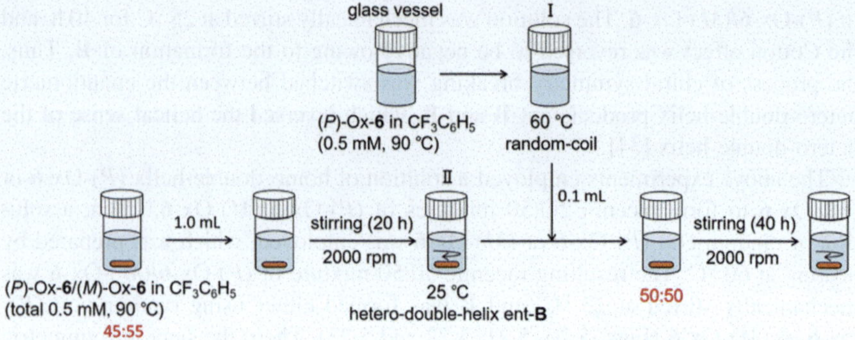

Fig. 5.22 Procedure of mixing experiment of hetero-double-helix **B** (45:55) and random-coil (*P*)-Ox-**6**

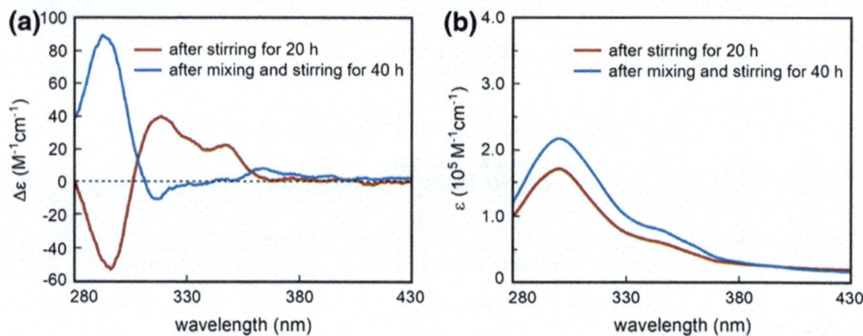

Fig. 5.23 CD spectra (**a**) and UV-vis spectra (**b**) of (*P*)-Ox-**6**/(*M*)-Ox-**6** in trifluoromethylbenzene (0.5 mM) at 25 °C formed by mixing experiment of hetero-double-helix ent-**B** (45:55) and random-coil (*P*)-Ox-**6**

(*P*)-Ox-**6** to (*P*)-Ox-**6**/(*M*)-Ox-**6** mixtures inverted ent-**B** to **B**, which indicated the presence of an equilibrium between **B**/ent-**B**.

It was also noted that the history of the preparation of racemic 50:50 mixtures of (*P*)-Ox-**6**/(*M*)-Ox-**6** switched the enantiomeric hetero-double-helix products **B** and ent-**B**. When a racemic 50:50 mixture was formed by mixing a stirred 45:55 mixture of (*P*)-Ox-**6**/(*M*)-Ox-**6** and a 55:45 mixture of (*P*)-Ox-**6**/(*M*)-Ox-**6**, ent-**B** was formed (Fig. 5.14c). When a racemic 50:50 mixture was formed by mixing a stirred 45:55 mixture of (*P*)-Ox-**6**/(*M*)-Ox-**6** and (*P*)-Ox-**6** or (*M*)-Ox-**6**, **B** was formed (Figs. 5.17, 5.19, 5.21 and 5.23). The unsymmetric nature can be seen by CD spectra. A slight deviation in the mixing procedures provides enantiomeric **B** and ent-**B**.

5.3 Mechanistic Aspects of Chiral Symmetry Breaking

The process of chiral symmetry breaking of (*P*)-Ox-**6**/(*M*)-Ox-**6** can be mechanically tuned, which exhibited notable property. The process was stopped and restarted by stopping and restarting mechanical stirring, which generated high-temperature domains on the solid surfaces owing to friction (Fig. 5.15). In addition, mixing procedures to form racemic 50:50 mixtures of (*P*)-Ox-**6**/(*M*)-Ox-**6** caused enantiomeric hetero-double-helix products **B** and ent-**B** to switch (Figs. 5.14, 5.17 and 5.19).

Chiral symmetry breaking of oxymethylenehelicene (*P*)-Ox-**6**/(*M*)-Ox-**6** and deterministic nature of aminomethylenehelicene oligomers [29] are compared (Fig. 5.2). The former exhibiting proximate stochastic chiral symmetry breaking involves slow hetero-double-helix formation, which is ascribed to higher energy barrier. The latter exhibiting deterministic chiral symmetry breaking involves faster hetero-double-helix formation, which is ascribed to lower energy barrier. Reaction media, in which chiral symmetry breaking occurs, are also different, and thermal activation by friction appears to be critical for (*P*)-Ox-**6**/(*M*)-Ox-**6**. The different nature of the helicene oligomers is due to the involvement of oxygen atoms and basic nitrogen atoms in the oligomer structures.

The different chiral symmetry breaking of oxymetylene and aminomethylene helicene oligomers may be due to different responses to chiral perturbations (Fig. 5.2), and following explanation is proposed. Chiral symmetry breaking of aminomethylene helicene oligomers [29] is much faster than oxymethylene oligomers. It suggests the higher acceleration efficiency of the former to subtle chiral perturbations resulting in deterministic chiral symmetry breaking. In contrast, slow process of oxymethylene oligomers involves lower acceleration by relatively strong chiral perturbations, resulting in proximate stochastic chiral symmetry breaking. It may be reasonable to consider that higher acceleration process is sensitive to deterministic subtle chiral perturbations and lower acceleration process responds to stochastic stronger chiral perturbations.

Chiral symmetry breaking of (*P*)-Ox-**6**/(*M*)-Ox-**6** can be described using a bifurcation model based on the concentration/extent of reaction profiles (Fig. 5.24). Stochastic chiral symmetry breaking is described by the symmetric nature of **B** or ent-**B** formation, which can be initiated from racemic 50:50 mixtures (Fig. 5.24a) or non-50:50 mixtures (Fig. 5.24b). Changes in conditions during the process are described by a bifurcation model, in which external perturbations to form **B** or ent-**B** override the effects of changes in conditions (Fig. 5.24c). For example, a 55:45 mixture of (*P*)-Ox-**6**/(*M*)-Ox-**6** provides **B**, to which a (*P*)-Ox-**6**/(*M*)-Ox-**6** mixture may be added to form a racemic 50:50 mixture, and mechanical stirring provides **B** (Fig. 5.14a). The enantiomeric reaction forms ent-**B** (Fig. 5.14c). These are symmetric with regard to chirality. In contrast, proximate stochastic chiral symmetry breaking (Fig. 5.11) is described in unsymmetric manner to form **B** or ent-**B**, which can be started from racemic 50:50 mixtures (Fig. 5.24d) or non-50:50 mixtures (Fig. 5.24e). Mechanical tuning of the process is described by a bifurcation model, in which external perturbations to form **B** override the effect of changes in conditions to form ent-**B**

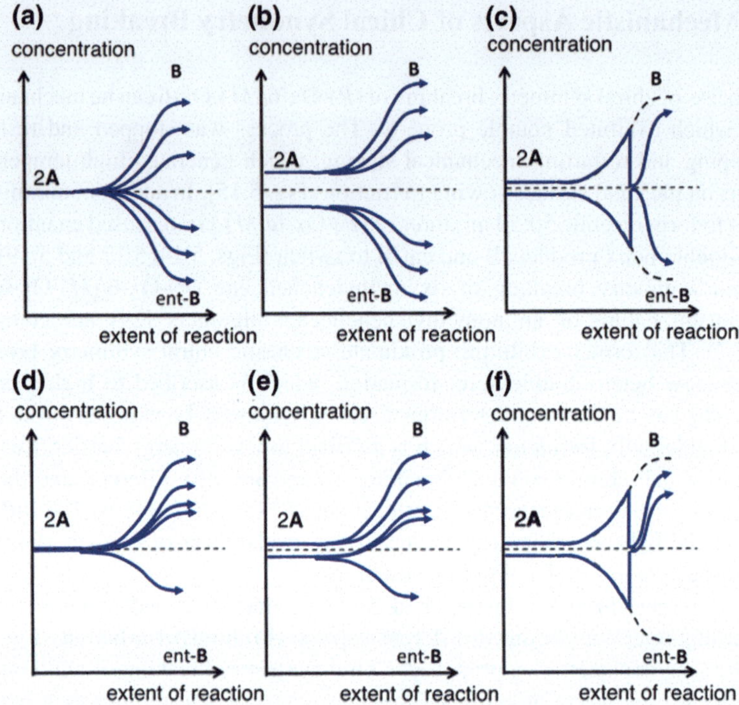

Fig. 5.24 Mechanistic model of chiral symmetry breaking described by concentration/extent of reaction profiles with bifurcation. Stochastic model initiated from **a** racemic 50:50 mixtures and **b** non-50:50 mixtures; **c** tuning of process. Proximate stochastic model initiated from **d** racemic 50:50 mixtures and **e** non-50:50 mixtures; **f** tuning of process. Horizontal dashed lines indicate racemic 50:50 mixtures

(Fig. 5.24f). For example, a 45:55 mixture of (*P*)-Ox-**6**/(*M*)-Ox-**6** provides ent-**B**, to which (*P*)-Ox-**6** is added to form a racemic 50:50 mixture, and mechanical stirring provides **B** (Fig. 5.19). The enantiomeric reaction initially forms **B**, and mechanical stirring provided **B** (Fig. 5.17). Slight differences in the procedures to prepare racemic 50:50 mixtures resulted in different products. These mechanical tuning phenomena are derived from the involvement of competitive self-catalytic reactions and their high sensitivity to external perturbations [37].

5.4 Summary

To summarize, oxymethylenehelicene (*P*)-Ox-**6**/(*M*)-Ox-**6** exhibited proximate stochastic chiral symmetry breaking in the formation of enantiomeric hetero-double-helices and aggregates, which was induced by mechanical stirring. Racemic 50:50 mixtures of (*P*)-Ox-**6**/(*M*)-Ox-**6** repeatedly provided hetero-double-helix **B**, and

46:54 and lower molar fractions of mixtures provided ent-**B**. The phenomenon is called proximate stochastic chiral symmetry breaking, which deviates slightly from symmetry. Random selection of **B** and ent-**B** occurs by small deviation from 50:50 mixture. The process of chiral symmetry breaking can be mechanically tuned by stopping and restarting mechanical stirring, which generates local and temporal heat to form hetero-double-helices. The process is also tuned by varying the mixing procedures for preparing racemic 50:50 mixtures. Mixing experiments using (*P*)-Ox-**6**/(*M*)-Ox-**6** mixtures revealed a stochastic nature, and mixing experiments using (*P*)-Ox-**6** alone revealed a proximate stochastic nature, which switched enantiomeric hetero-double-helix products ent-**B** and **B** by reversing the helical sense. The mechanical tuning phenomena are derived from competitive self-catalytic reactions and their high sensitivity to external perturbations.

References

1. Frank FC (1953) On spontaneous asymmetric synthesis. Biochim Biophys Acta 11:459
2. Kondepudi DK, Nelson GW (1983) Chiral Symmetry Breaking in Nonequilibrium Systems. Phys Rev Lett 50:1023
3. Kondepudi DK, Nelson GW (1984) Chiral-symmetry-breaking states and their sensitivity in nonequilibrium chemical systems. Physica 125A:465
4. Tranter GE (1985) The parity-violating energy differences between the enantiomers of α-amino acids. Chem Phys Lett 120:93
5. Mason SF, Tranter GE (1985) The electroweak origin of biomolecular handedness. Proc R Soc Lond A397:45
6. Kondepudi DK, Nelson GW (1985) Weak neutral currents and the origin of biomolecular chirality. Nature 314:438
7. Kondepudi DK (1987) Selection of molecular chirality by extremely weak chiral interactions under far-from-equilibrium conditions. Biosystems 20:75
8. Kondepudi DK, Asakura K (2001) Chiral Autocatalysis, Spontaneous Symmetry Breaking, and Stochastic Behavior. Acc Chem Res 34:946
9. Atkins P, de Paula J (2014) Physical chemistry, 10th edn. Oxford University Press, Oxford
10. Houston PL (2006) Chemical kinetics and reaction dynamics. Dover Publications Inc., New York
11. van Santen RA, Niemantsverdriet JW (1995) Chemical kinetics and catalysis. Plenum Press, New York
12. Kondepudi DK, Kaufman RJ, Singh N (1990) Chiral Symmetry Breaking in Sodium Chlorate Crystallizaton. Science 250:975
13. Viedma C (2005) Chiral Symmetry Breaking During Crystallization: Complete Chiral Purity Induced by Nonlinear Autocatalysis and Recycling. Phys Rev Lett 94:065504
14. Viedma C (2007) Selective Chiral Symmetry Breaking during Crystallization: Parity Violation or Cryptochiral Environment in Control? Cryst Growth Des 7:553
15. El-Hachemi Z, Crusats J, Ribó JM, McBride JM, Veintemillas-Verdaguer S (2011) Metastability in Supersaturated Solution and Transition towards Chirality in the Crystallization of $NaClO_3$. Angew Chem 123:2407; (2011) Angew Chem Int Ed 50:2359
16. Durand DJ, Kondepudi DK, Moreira PF Jr, Quina FH (2002) Generation of molecular chiral asymmetry through stirred crystallization. Chirality 14:284
17. McLaughlin DT, Nguyen TPT, Mengnjo L, Bian C, Leung YH, Goodfellow E, Ramrup P, Woo S, Cuccia LA (2014) Viedma Ripening of Conglomerate Crystals of Achiral Molecules Monitored Using Solid-State Circular Dichroism. Cryst Growth Des 14:1067

18. Tarasevych AV, Sorochinsky AE, Kukhar VP, Toupet L, Crassous J, Guillemin JC (2015) Attrition-induced spontaneous chiral amplification of the γ polymorphic modification of glycine. CrystEngComm 17:1513
19. Noorduin WL, Izumi T, Millemaggi A, Leeman M, Meekes H, van Enckevort WJP, Kellogg RM, Kaptein B, Vlieg E, Blackmond DG (2008) Emergence of a Single Solid Chiral State from a Nearly Racemic Amino Acid Derivative. J Am Chem Soc 130:1158
20. Noorduin WL, Meekes H, van Enckevort WJP, Kaptein B, Kellogg RM, Vlieg E (2010) The Driving Mechanism Behind Attrition-Enhanced Deracemization. Angew Chem 122:2593; (2010) Angew Chem Int Ed 49:2539
21. Yagishita F, Ishikawa H, Onuki T, Hachiya S, Mino T, Sakamoto M (2012) Angew Chem 124:13200; (2012) Total Spontaneous Resolution by Deracemization of Isoindolinones. Angew Chem Int Ed 51:13023
22. Yagishita F, Okamoto K, Kamataki N, Kanno S, Mino T, Kasashima Y, Sakamoto M (2013) Chiral Symmetry Breaking of Axially Chiral Nicotinamide by Crystallization from the Melt. Chem Lett 42:1508
23. Suwannasang K, Flood AE, Rougeot C, Coquerel G (2013) Using Programmed Heating–Cooling Cycles with Racemization in Solution for Complete Symmetry Breaking of a Conglomerate Forming System. Cryst Growth Des 13:3498
24. Kimura M, Hatanaka T, Nomoto H, Takizawa J, Fukawa T, Tatewaki Y, Shirai H (2010) Self-Assembled Helical Nanofibers Made of Achiral Molecular Disks Having Molecular Adapter. Chem Mater 22:5732
25. Wang Y, Zhou D, Li H, Li R, Zhong Y, Sun X, Sun X (2014) Hydrogen-bonded supercoil self-assembly from achiral molecular components with light-driven supramolecular chirality. J Mater Chem C 2:6402
26. Kirstein S, von Berlepsch H, Böttcher C, Burger C, Quart A, Reck G, Dähne S (2000) Chiral J-Aggregates Formed by Achiral Cyanine Dyes. ChemPhysChem 3:146
27. Azeroual S, Surprenant J, Lazzara TD, Kocun M, Tao Y, Cuccia LA, Lehn JM (2012) Mirror symmetry breaking and chiral amplification in foldamer-based supramolecular helical aggregates. Chem Commun 48:2292
28. Stals PJM, Korevaar PA, Gillissen MAJ, de Greef TFA, Fitié CFC, Sijbesma RP, Palmans ARA, Meijer EW (2012) Symmetry Breaking in the Self-Assembly of Partially Fluorinated Benzene-1,3,5-tricarboxamides. Angew Chem 124:11459; (2012) Angew Chem Int Ed 51:11297
29. Kushida Y, Sawato T, Shigeno M, Saito N, Yamaguchi M (2017) Deterministic and Stochastic Chiral Symmetry Breaking Exhibited by Racemic Aminomethylenehelicene Oligomers. Chem Eur J 23:327
30. Demirel Y (2007) Non-equilibrium thermodynamics: transport and rate processes in physical, chemical and biological systems, 2nd edn. Elsevier, Amsterdam
31. Epstein IR, Pojman JA (1998) An introduction to nonlinear chemical dynamics. Oxford University Press, Oxford
32. Strogatzs SH (1994) Nonlinear dynamics and chaos. Perseus Books Publishing
33. de Groot SR, Mazur P (1984) Non-equilibrium thermodynamics. Dover Publication, New York
34. Shigeno M, Kushida Y, Yamaguchi M (2014) Heating/Cooling Stimulus Induces Three-State Molecular Switching of Pseudoenantiomeric Aminomethylenehelicene Oligomers: Reversible Nonequilibrium Thermodynamic Processes. J Am Chem Soc 136:7972
35. Saito N, Yamaguchi M (2018) Synthesis and Self-Assembly of Chiral Cylindrical Molecular Complexes: Functional Heterogeneous Liquid-Solid Materials Formed by Helicene Oligomers. Molecules 23:277
36. Sawato T, Saito N, Shigeno M, Yamaguchi M (2017) Mechanical Stirring Induces Heteroaggregate Formation and Self-assembly of Pseudoenantiomeric Oxymethylene Helicene Oligomers in Solution. ChemistrySelect 2:2205
37. Sawato T, Saito N, Yamaguchi M (2019) Chemical Systems Involving Two Competitive Self-Catalytic Reactions. ACS Omega 4:5879
38. Shigeno M, Sawato T, Yamaguchi M (2015) Fibril Film Formation of Pseudoenantiomeric Oxymethylenehelicene Oligomers at the Liquid–Solid Interface: Structural Changes, Aggregation, and Discontinuous Heterogeneous Nucleation. Chem Eur J 21:17676

Chapter 6
Conclusions

Abstract The conclusion of each chapter is described.

Keywords Liquid–solid interface · Helicene · Hetero-double-helix · Mechanical stirring · Chiral symmetry breaking

1. Oxymethylenehelicene (*P*)- and (*M*)-oligomers up to nonamer were synthesized by a building block method, which formed homo-double-helix in trifluoromethylbenzene.
2. A mixture of the pseudoenantiomeric (*P*)-pentamer and (*M*)-hexamer formed a hetero-double-helix, which self-assembled into fibril films at the liquid–solid interface. Discontinuous heterogeneous nucleation occurred, which involved the formation of 50-nm-diameter particles and subsequent fibril growth from these particles. The fibril film was formed on the solid surface, and the molecules remained dissociated in solution. The fibril film formation was affected by seeding and the solid surface materials.
3. A mixture of pseudoenantiomeric oxymethylene helicene (*P*)-pentamers and (*M*)-hexamers formed hetero-double-helix and self-assembled as a result of mechanical stirring in solution at 25 °C. No aggregation occurred without stirring, and hetero-double-helix formation was accelerated at high stirring rates. Repeatedly alternating between mechanical stirring and stopping respectively induced and then ceased the hetero-double-helix formation. Sonication also promoted the hetero-double-helix formation, which did not occur at high pressures or during centrifugation. Mechanisms based on friction due to mechanical stirring, which generated local and temporal high-temperature domains, are considered. It is shown that different methods of mechanical stimulation, namely, stirring and surface contact, induce enantiomeric hetero-double-helix formation.
4. A racemic 50:50 mixture of oxymethylenehelicene (*P*)-hexamer and (*M*)-hexamer in solution exhibits proximate stochastic chiral symmetry breaking during formation of enantiomeric hetero-double-helices and their aggregates which was induced by mechanical stirring. Mechanical stirring with a magnetic stirrer at 2000 rpm for 100 h at 25 °C resulted in the exhibition of a strong negative Cotton

T. Sawato, *Synthesis of Optically Active Oxymethylenehelicene Oligomers and Self-assembly Phenomena at a Liquid–Solid Interface*, Springer Theses,
https://doi.org/10.1007/978-981-15-3192-7_6

effect at 322 nm, which is chiral symmetry breaking. No change in the Cotton effect occurred in the absence of stirring. The (*P*)-hexamer to (*M*)-hexamer mixing molar fraction was varied, and a positive Cotton effect appeared at molar fractions between 40:60 and 46:54 and a negative Cotton effect at molar fractions between 48:52 and 60:40, which was reversed at 47:53 with slight deviation of symmetry from that at 50:50. The phenomenon was termed proximate stochastic chiral symmetry breaking. The process of chiral symmetry breaking could be tuned by varying the procedures of mechanical stirring and mixing procedures for solutions of (*P*)-hexamer and/or (*M*)-hexamer.

Chapter 7
Experimental Section

Abstract General and detailed experimental methods are described.

Keywords Circular dichroism spectroscopy · Ultraviolet–visible absorption spectroscopy · Atomic force microscope · Dynamic light scattering · Vapor pressure osmometry

7.1 General Methods

All reactions were carried out under an argon atmosphere. Column chromatography was carried out on silica gel 60 N (40–50 μm). NMR spectra were measured on a Varian Mercury (^{1}H at 400 MHz and ^{13}C at 100 MHz). ^{1}H NMR taken in $CDCl_3$ (δ 7.26) were referenced to the residual solvent. ^{13}C NMR spectra taken in $CDCl_3$ (δ 77.0) was referenced to the residual solvent. Melting points were determined with a Yanagimoto micro melting point apparatus without correction. Elemental analyses were conducted with Yanaco CHN CORDER MT-6. Optical rotations were measured on a JASCO P-1010 digital polarimeter. IR spectra were measured on a JASCO FT/IR-400 spectrophotometer. Low- and high-resolution mass spectra were recorded on a JEOL JMS T-100GC, or a JEOL JMS-700 spectrometer. MALDI TOF-MS spectra were recorded on a Shimadzu Axima spectrometer observing with positive ion mode and using α-cyano-4-hydroxycinnamic acid as a matrix. FAB mass spectra were recorded on a JEOL JMS-700 spectrometer by using 3-nitrobenzyl alcohol (NBA) matrix. CD spectra were measured on a JASCO J-720 spectropolarimeter. Vapor pressure osmometry (VPO) was conducted with a KNAUER K-7000 molecular weight apparatus using benzyl as a standard. DLS results were determined at 173° scattering angle by a Zetasizer Nano S90. Atomic force microscopic (AFM) images were obtained on a Veeco Digital Instrument Multimode Nanoscope IIIa in the tapping mode regime under ambient temperature and pressure. Micro-fabricated silicon cantilever tips (Olympus OMCL-AC160TS-C2) were used. Polarized optical microscopic observation (POM) were carried out with an Olympus BX53 optical polarizing microscope equipped with a microscope digital camera Olympus DP21, a Linkam 10083 hot stage, and a freezing stage. Gel permeation chromatography (GPC) was conducted with a Recycling Preparative HPLC LC-908 or LC-918 (Japan Analytical Industry, Co. Ltd.).

T. Sawato, *Synthesis of Optically Active Oxymethylenehelicene Oligomers and Self-assembly Phenomena at a Liquid–Solid Interface*, Springer Theses,
https://doi.org/10.1007/978-981-15-3192-7_7

7.2 Experimental Method for Chap. 2

(*P*)-5,8-Bis(bromomethyl)-1,12-dimethylbenzo[*c*]phenanthrene, (*P*)-12. Under an argon atmosphere, to a solution of (*P*)-**11** (316 mg, 1.00 mmol) in dichloromethane (8 ml), a solution of tribromophosphine (105 μL, 1.12 mmol) in dichloromethane (8 ml) was added at 0 °C, and the mixture was stirred at room temperature for 2 h. The reaction was quenched by adding saturated aqueous sodium hydrogen carbonate, and the organic materials were extracted with dichloromethane three times. The combined organic layers were washed with water and brine, and dried over sodium sulfate. The solvent was evaporated under reduced pressure, and purification by silica gel chromatography (toluene) gave (*P*)-**12** (319 mg, 0.721 mmol, 72%). MS (EI) *m/z* Calcd. for $C_{22}H_{18}Br_2$ (M^+): 439.9775. Found: 439.9758. Anal. ($C_{22}H_{18}Br_2$) Calcd: C 59.76, H 4.10. Found: C 59.48, H 4.20. $[\alpha]^{27}{}_D$ −59.1 (c = 1.0, $CHCl_3$). IR (KBr) 2952, 2924, 671 cm^{-1}. ^1H NMR (400 MHz, $CDCl_3$) δ 1.93 (6H, s), 5.03 (2H, d, *J* = 10.4 Hz), 5.15 (2H, d, *J* = 10.4 Hz), 7.44 (2H, d, *J* = 7.2 Hz), 7.68 (2H, t, *J* = 7.6 Hz), 7.82 (2H, s), 8.18 (2H, d, *J* = 8.0 Hz). ^{13}C NMR (100 MHz, $CDCl_3$), δ 23.5, 31.8, 121.08, 126.4, 126.8, 127.5, 128.8, 130.2, 131.0, 131.9, 132.4, 137.2.

Decyl 3-Hydroxy-{5-(tetrahydro-2*H*-pyran-2-yl)oxy}benzoate 13. Under an argon atmosphere, to a solution of **14** (4.00 g, 13.6 mmol) in dichloromethane (40 ml), pyridinium *p*-toluenesulfonate (341 mg, 1.36 mmol) and 3,4-Dihydro-2*H*-pyran (1.25 ml, 13.7 mmol) was added, and the mixture was stirred at room temperature for 2 h. The reaction was quenched by adding saturated aqueous sodium hydrogen carbonate, and the organic materials were extracted with dichloromethane three times. The combined organic layers were washed with water and brine, and dried over sodium sulfate. The solvent was evaporated under reduced pressure, and purification by silica gel chromatography (hexane:ethyl acetate = 16:1, 8:1, and 4:1) gave **13** (2.73 g, 7.20 mmol, 53%).MS (EI) *m/z* Calcd. for $C_{22}H_{34}O_5$ (M^+): 378.2406. Found: 378.2434. Anal. ($C_{22}H_{34}O_5$) Calcd: C 69.81, H 9.05. Found: C 69.85, H 8.99. IR (neat) 3391, 2926, 2854, 1720, 1694, 1163 cm^{-1}. ^1H NMR (400 MHz, $CDCl_3$) δ 0.88 (3H, t, *J* = 6.8 Hz), 1.24–1.32 (14H, m), 1.43 (2H, br), 1.63 (2H, m), 1.74 (2H, quint, *J* = 7.6 Hz), 1.85 (2H, quint, *J* = 4.4 Hz), 3.60 (1H, d, *J* = 11.6 Hz), 3.89 (1H, t, *J* = 8.8 Hz), 4.28 (2H, t, *J* = 6.4 Hz), 5.42 (1H, t, *J* = 2.8 Hz), 6.82 (1H, t, *J* = 1.2 Hz), 7.22 (1H, t, *J* = 1.2 Hz), 7.26 (1H, t, *J* = 7.2 Hz), 7.69 (1H, s). ^{13}C NMR (100 MHz, $CDCl_3$), δ 13.9, 18.5, 20.8, 22.5, 24.9, 25.8, 28.4, 29.1, 29.3, 30.0, 31.7, 60.4, 61.8, 65.2, 96.2, 108.4, 109.4, 109.8, 157.4, 157.8, 166.6, 171.5.

(*P*)-5,8-Bis[5-(decyloxycarbonyl)-{3-(tetrahydro-2*H*-pyran-2-yl)oxy}phenoxymethyl]-1,12-dimethylbenzo[*c*]phenanthrene, (*P*)-Ox-1. Under an argon atmosphere, to a solution of (*P*)-**12** (43.8 mg, 0.099 mmol) and **13** (79.9 mg, 0.211 mmol) in *N,N*-dimetylformamide (2 ml), sodium hydride (8.3 mg, 0.208 mmol) was added at 0 °C, and the mixture was stirred at room temperature for 2 h. The reaction was quenched by adding water, and the organic materials were extracted with ethyl acetate three times. The combined organic layers were washed with water and brine, and dried over sodium sulfate. The solvent was evaporated under reduced pressure, and purification by silica gel chromatography (hexane:ethyl acetate = 8:1) and GPC (toluene) gave (*P*)-Ox-**1** (90.2 mg, 0.087 mmol, 88%). MS (FAB, NBA) *m/z* Calcd. for $C_{66}H_{84}O_{10}$ ($[M]^+$): 1036.6065. Found: 1036.6020. Anal. ($C_{66}H_{84}O_{10}$) Calcd: C 76.42, H 8.16. Found: C 76.37, H 8.25. Mp 81–83 °C (CH_2Cl_2-hexane). $[\alpha]^{27}{}_D$ −21.1 (c = 1.0, $CHCl_3$). IR (KBr) 2925, 2853, 1718, 1594, 1162 cm^{-1}. ^1H NMR (400 MHz, $CDCl_3$) δ 0.86 (6H, t, *J* = 6.8 Hz), 1.24–1.33 (26H, m), 1.42 (4H, br), 1.67 (6H, m), 1.75 (4H, quint, *J* = 7.6 Hz), 1.86 (4H, quint, *J* = 4.4 Hz), 1.98 (6H, s), 3.60 (2H, d, *J* = 11.6 Hz), 3.89 (2H, t, *J* = 8.8 Hz), 4.30 (4H, t, *J* = 6.4 Hz), 5.46 (2H, t, *J* = 2.8 Hz), 5.64 (4H, q, *J* = 12.0 Hz), 7.03 (2H, s), 7.40 (2H, s), 7.44 (2H, d, *J* = 7.2 Hz), 7.47 (2H, s), 7.62 (2H, t, *J* = 7.6 Hz), 7.94 (2H, s), 8.05 (2H, d, *J* = 8.0 Hz). ^{13}C NMR (100 MHz, $CDCl_3$), δ 14.1, 18.7, 22.7, 23.5, 25.1, 26.0, 28.7, 29.27, 29.29, 29.5, 30.2, 31.9, 62.2, 65.3, 68.9, 96.6, 108.5, 108.8, 110.8, 120.9, 126.0, 126.3, 126.5, 128.4, 130.7, 131.2, 131.7, 132.5, 137.1, 158.1, 159.8, 166.4.

(*P*)-8-Bromomethyl-5-[5-(decyloxycarbonyl)-{3-(tetrahydro-2*H*-pyran-2-yl)oxy}phenoxymethyl]-1,12-dimethylbenzo[*c*]phenanthrene, (*P*)-10. Under an argon atmosphere, to a solution of (*P*)-**12** (209 mg, 0.473 mmol) and **13** (179 mg, 0.473 mmol) in *N,N*-dimetylformamide (10.5 ml), sodium hydride (11.3 mg, 0.473 mmol) was added at 0 °C, and the mixture was stirred at room temperature for 1 h. The reaction was quenched by adding water, and the organic materials were extracted with ethyl acetate three times. The combined organic layers were washed with water and brine, and dried over sodium sulfate. The solvent was evaporated under reduced pressure, and purification by silica gel chromatography (hexane:ethyl acetate = 20:1 and 10:1) and GPC (THF) gave (*P*)-**10** (125 mg, 0.169 mmol, 36%). MS (FAB, NBA) *m/z* Calcd. for $C_{44}H_{51}BrO_5$ (M^+): 738.2920. Found: 738.2876. $[\alpha]^{27}{}_D$ −37.0 (c = 1.0, $CHCl_3$). IR (neat) 2925, 2854, 1715, 1593, 1160, 681 cm^{-1}. ^1H NMR (400 MHz, $CDCl_3$) δ 0.87 (3H, t, *J* = 6.8 Hz), 1.26–1.33 (13H, m), 1.42 (2H, br), 1.65 (3H, m), 1.75 (2H, quint, *J* = 7.6 Hz), 1.85 (2H, quint, *J* = 4.4 Hz), 1.93 (6H, s), 3.60 (1H, d, *J* = 11.6 Hz), 3.88 (1H, t, *J* = 8.8 Hz), 4.30 (2H, t, *J* = 6.4 Hz), 5.01 (1H, d, *J* = 10.4 Hz), 5.13 (1H, d, *J* = 10.4 Hz), 5.46 (1H, t, *J* = 2.8 Hz), 5.64 (2H, q, *J* = 12.0 Hz), 7.02 (1H, s), 7.40 (2H, s), 7.43 (1H, s), 7.47 (1H, s), 7.61 (1H, t, *J* = 7.2 Hz), 7.66 (1H, t, *J* = 7.2 Hz), 7.85 (2H, d, *J* = 8.4 Hz), 8.03 (1H, d, *J* = 8.4 Hz), 8.17 (1H, d, *J* = 8.4 Hz). ^{13}C NMR (100 MHz, $CDCl_3$), δ 14.1, 18.7, 22.6, 23.5, 25.1, 26.0, 28.7, 29.3, 29.5, 30.2, 31.8, 32.0, 62.1, 65.3, 68.7, 96.4, 108.4, 108.5, 108.7, 110.7, 121.0, 125.5, 126.2, 126.4, 126.9, 127.1, 128.5, 128.6, 130.0, 130.7, 131.0, 131.3, 131.5, 131.9, 132.1, 132.4, 137.0, 158.1, 159.7, 166.3.

Deprotected monomer (*P*)-Ox-1H. Under an argon atmosphere, to a solution of (*P*)-Ox-**1** (241.8 mg, 0.233 mmol) in dichloromethane/methanol (1:1, 16.3 mL), a catalytic amount of *p*-toluenesulfonic acid monohydrate (44.3 mg, 0.233 mmol) was added, and the mixture was stirred at room temperature for 3 h. The reaction was quenched by adding water, and the organic materials were extracted with dichloromethane three times. The combined organic layers were washed with water, brine, and dried over sodium sulfate. The solvent was evaporated under reduced pressure, and purification by silica gel chromatography (toluene:ethyl acetate = 20:1, 12:1 and 8:1) and GPC (toluene) gave (*P*)-Ox-**1H** (180.6 mg, 0.208 mmol, 89%). MS (MALDI-TOF) *m/z* Calcd. for $C_{56}H_{68}NaO_8$ ($[M+Na]^+$): 891.48. Found: 891.95. Anal. ($C_{56}H_{68}O_8$) Calcd: C 77.39, H 7.89. Found: C 77.36, H 7.84. Mp 130–132 °C (CH_2Cl_2-hexane). $[\alpha]^{27}{}_D$ −21.8 (c = 1.0, $CHCl_3$). IR (KBr) 3355, 2952, 2921, 2850, 1719, 1682, 1160 cm^{-1}. ^1H NMR (400 MHz, $CDCl_3$) δ 0.86 (6H, t, *J* = 6.8 Hz), 1.25–1.31 (24H, m), 1.40 (4H, br), 1.73 (4H, quint, *J* = 7.2 Hz), 1.94 (6H, s), 4.29 (4H, t, *J* = 6.8 Hz), 5.55 (4H, q, *J* = 12.8 Hz), 5.64 (2H, s), 6.80 (2H, s), 7.22 (2H, s), 7.39 (2H, s), 7.41 (2H, d, *J* = 10.0 Hz), 7.59 (2H, t, *J* = 7.6 Hz), 7.83 (2H, s), 7.99 (2H, d, *J* = 8.4 Hz). ^{13}C NMR (100 MHz, $CDCl_3$), δ 14.1, 22.7, 23.5, 26.0, 28.7, 29.26, 29.28, 29.51, 29.53, 31.9, 65.6, 68.9, 107.4, 108.3, 109.7, 120.8, 126.0, 126.3, 126.4, 128.4, 130.6, 130.98, 131.01, 131.6, 132.6, 137.1, 156.9, 160.0, 166.6.

Dimer (*M*)-Ox-2. Under an argon atmosphere, to **14** (56.0 mg, 0.190 mmol), a solution of (*M*)-**10** (281.6 mg, 0.380 mmol) in *N,N*-dimethylformamide (9.0 mL) and sodium hydride (10.0 mg, 0.419 mmol) was added at 0 °C, and the mixture was stirred at room temperature for 3 h. The reaction was quenched by adding water, and the organic materials were extracted with ethyl acetate three times. The combined organic layers were washed with water and brine, and dried over sodium sulfate. The solvent was evaporated under reduced pressure, and purification by silica gel chromatography (hexane:ethyl acetate = 10:1 and 4:1) and GPC (toluene) gave (*M*)-Ox-**2** (278.7 mg, 0.173 mmol, 93%). MS (MALDI-TOF) *m/z* Calcd. for $C_{105}H_{126}NaO_{14}$ ($[M+Na]^+$): 1633.90. Found: 1633.90. IR (neat) 2926, 2853, 1718, 1685, 1160 cm^{-1}. ^1H NMR (400 MHz, $CDCl_3$) δ 0.86 (9H, t, *J* = 6.8 Hz), 1.24–1.33 (34H, m), 1.42 (6H, br), 1.67 (8H, m), 1.75 (6H, quint, *J* = 7.6 Hz), 1.86 (6H, quint, *J* = 4.4 Hz), 1.98 (12H, s), 3.60 (2H, d, *J* = 11.6 Hz), 3.89 (2H, t, *J* = 8.8 Hz), 4.30 (6H, t, *J* = 6.4 Hz), 5.46 (2H, t, *J* = 2.8 Hz), 5.64 (8H, q, *J* = 12.0 Hz), 7.03 (3H, s), 7.40 (3H, s), 7.44 (4H, d, *J* = 7.2 Hz), 7.47 (3H, s), 7.62 (4H, t, *J* = 7.6 Hz), 7.94 (4H, s), 8.05 (4H, d, *J* = 8.0 Hz). ^{13}C NMR (100 MHz, $CDCl_3$), δ 14.1, 18.7, 22.7, 23.5, 25.1, 26.0, 28.7, 29.26, 29.28, 29.5, 30.2, 31.9, 62.1, 65.3, 68.9, 96.5, 108.5, 108.8, 110.8, 120.9, 126.0, 126.3, 126.4, 128.4, 130.7, 131.1, 131.6, 132.5, 137.0, 158.1, 159.8, 166.4.

Deprotected dimer (*M*)-Ox-2H. Under an argon atmosphere, to a solution of (*M*)-Ox-**2** (238.1 mg, 0.148 mmol) in dichloromethane/methanol (1:1, 10.8 mL), a catalytic amount of *p*-toluenesulfonic acid monohydrate (28.1 mg, 0.148 mmol) was added, and the mixture was stirred at room temperature for 2 h. The reaction was quenched by adding water, and the organic materials were extracted with dichloromethane three times. The combined organic layers were washed with water, brine, and dried over sodium sulfate. The solvent was evaporated under reduced pressure, and purification by silica gel chromatography (hexane:ethyl acetate = 4:1 and 2:1) and GPC (toluene) gave (*M*)-Ox-**2H** (202.8 mg, 0.141 mmol, 95%). MS (MALDI-TOF) *m/z* Calcd. for $C_{95}H_{110}NaO_{12}$ ($[M+Na]^+$): 1465.79. Found: 1465.59. Anal. ($C_{95}H_{110}O_{12}$) Calcd: C 79.02, H 7.68. Found: C 78.96, H 7.73 Mp 71–73 °C

(CH_2Cl_2-hexane). $[\alpha]^{27}_D$ 31.3 (c = 1.0, $CHCl_3$). IR (KBr) 3392, 2925, 2853, 1717, 1697, 1160 cm^{-1}. ^{1}H NMR (400 MHz, $CDCl_3$) δ 0.86 (9H, t, J = 6.8 Hz), 1.25–1.31 (36H, m), 1.40 (6H, br), 1.73 (6H, quint, J = 7.2 Hz), 1.93 (12H, s), 4.29 (6H, t, J = 6.8 Hz), 5.56 (8H, q, J = 12.8 Hz), 6.48 (2H, s), 6.80 (2H, s), 7.02 (1H, s), 7.23 (2H, s), 7.37 (2H, s), 7.40 (4H, d, J = 10.0 Hz), 7.48 (2H, s), 7.58 (4H, t, J = 7.6 Hz), 7.85 (4H, s), 8.00 (4H, d, J = 8.4 Hz). ^{13}C NMR (100 MHz, $CDCl_3$), δ 14.1, 22.6, 23.5, 26.0, 28.59, 28.64, 29.3, 29.49, 29.51, 31.8, 65.6, 68.8, 68.9, 107.3, 108.1, 108.7, 109.8, 120.78, 120.84, 125.9, 126.1, 126.2, 126.3, 128.4, 130.5, 130.6, 130.92, 130.94, 131.0, 131.5, 131.6, 132.4, 132.5, 137.0, 157.1, 159.9, 160.0, 166.7.

Trimer (*P*)-Ox-3. Under an argon atmosphere, to (*P*)-Ox-**1H** (58.9 mg, 67.8 μmol), a solution of (*P*)-**10** (125.4 mg, 0.170 mmol) in *N*,*N*-dimethylformamide (8.0 mL) and sodium hydride (4.1 mg, 0.170 mmol) was added at 0 °C, and the mixture was stirred at room temperature for 3 h. The reaction was quenched by adding water, and the organic materials were extracted with ethyl acetate three times. The combined organic layers were washed with water and brine, and dried over sodium sulfate. The solvent was evaporated under reduced pressure, and purification by silica gel chromatography (hexane:ethyl acetate = 12:1, 8:1 and 4:1) and GPC (toluene) gave (*P*)-Ox-**3** (131.9 mg, 60.3 μmol, 89%). MS (MALDI-TOF) *m/z* Calcd. for $C_{144}H_{168}NaO_{18}$ ($[M+Na]^+$): 2208.21. Found: 2208.80. Anal. ($C_{144}H_{168}O_{18}$) Calcd: C 79.09, H 7.74. Found: C 78.87, H 7.86. Mp 76–78 °C (CH_2Cl_2-hexane). $[\alpha]^{27}_D$ −25.9 (c = 1.0, $CHCl_3$). IR (KBr) 2925, 2853, 1719, 1595, 1159 cm^{-1}. ^{1}H NMR (400 MHz, $CDCl_3$) δ 0.85 (12H, t, J = 6.8 Hz), 1.25–1.32 (40H, m), 1.42 (8H, br), 1.65 (12H, m), 1.75 (8H, quint, J = 7.6 Hz), 1.85 (8H, quint, J = 4.4 Hz), 1.96 (18H, s), 3.59 (2H, d, J = 11.6 Hz), 3.88 (2H, t, J = 8.8 Hz), 4.31 (8H, t, J = 6.4 Hz), 5.45 (2H, t, J = 2.8 Hz), 5.63 (12H, q, J = 12.0 Hz), 7.03 (2H, s), 7.05 (2H, s), 7.40 (2H, s), 7.42 (6H, d, J = 7.2 Hz), 7.47 (2H, s), 7.51 (4H, s), 7.60 (6H, t, J = 7.6 Hz), 7.93 (6H, s), 8.04 (6H, d, J = 8.0 Hz). ^{13}C NMR (100 MHz, $CDCl_3$), δ 14.1, 18.7, 22.6, 23.5, 25.1, 26.0, 28.7, 29.3, 29.5, 30.2, 31.9, 62.1, 65.3, 65.4, 68.9, 69.0, 96.5, 107.1, 108.5, 108.7, 110.8, 120.9, 126.0, 126.1, 126.3, 126.48, 126.54, 128.4, 130.65, 130.68, 131.01, 131.03, 131.1, 131.2, 131.6, 132.5, 132.7, 137.0, 137.1, 158.1, 159.8, 160.0, 166.35, 166.37.

Deprotected trimer (*P*)-Ox-3H. Under an argon atmosphere, to a solution of (*P*)-Ox-**3** (180.4 mg, 82.5 μmol) in dichloromethane/methanol (1:1, 6.0 mL), a catalytic amount of *p*-toluenesulfonic acid monohydrate (15.7 mg, 82.5 μmol) was added, and the mixture was stirred at room temperature for 2 h. The reaction was quenched by adding water, and the organic materials were extracted with dichloromethane three times. The combined organic layers were washed with water, brine, and dried over sodium sulfate. The solvent was evaporated under reduced pressure, and purification by silica gel chromatography (hexane:ethyl acetate = 4:1 and 2:1) and GPC (toluene) gave (*P*)-Ox-**3H** (163.7 mg, 81.1 μmol, 98%). MS (MALDI-TOF) *m/z* Calcd. for $C_{134}H_{152}NaO_{16}$ ($[M+Na]^+$): 2040.10. Found: 2039.88. Anal. ($C_{134}H_{152}O_{16}$) Calcd: C 79.73, H 7.59. Found: C 79.52, H 7.51 Mp 92–94 °C (CH_2Cl_2-hexane). $[\alpha]^{27}_D$ −31.6 (c = 0.10, $CHCl_3$). IR (KBr) 3420, 2925, 2853, 1717, 1697, 1158 cm^{-1}. ^{1}H

NMR (400 MHz, $CDCl_3$) δ 0.84 (12H, t, J = 6.8 Hz), 1.24–1.31 (50H, m), 1.39 (8H, br), 1.73 (8H, quint, J = 7.2 Hz), 1.94 (18H, s), 4.28 (8H, t, J = 6.8 Hz), 5.58 (12H, q, J = 12.8 Hz), 6.79 (2H, s), 7.02 (2H, s), 7.19 (2H, s), 7.38 (2H, s), 7.40 (6H, d, J = 10.0 Hz), 7.48 (4H, s), 7.58 (6H, t, J = 7.6 Hz), 7.86 (6H, s), 8.00 (6H, d, J = 8.4 Hz). ^{13}C NMR (100 MHz, $CDCl_3$), δ 14.1, 22.6, 23.5, 26.0, 28.6, 28.7, 29.3, 29.50, 29.52, 31.9, 65.5, 65.6, 68.86, 68.94, 107.3, 108.1, 108.7, 109.7, 120.8, 125.9, 126.0, 126.1, 126.3, 126.4, 128.4, 130.6, 130.98, 131.01, 131.6, 132.55, 132.58, 137.0, 157.0, 159.9, 160.0, 166.5, 166.7.

Tetramer (*M*)-Ox-4. Under an argon atmosphere, to (*M*)-Ox-**2H** (256.6 mg, 0.178 mmol), a solution of (*M*)-**10** (433.9 mg, 0.587 mmol) in *N,N*-dimethylformamide (35.0 mL) and potassium carbonate (98.2 mg, 0.711 mmol) was added at 0 °C, and the mixture was stirred at 35 °C for 4 h. The reaction was quenched by adding water, and the organic materials were extracted with ethyl acetate three times. The combined organic layers were washed with water and brine, and dried over sodium sulfate. The solvent was evaporated under reduced pressure, and purification by silica gel chromatography (hexane:ethyl acetate = 8:1 and 4:1) and GPC (toluene) gave (*M*)-Ox-**4** (396.9 mg, 0.144 mmol, 81%). MS (MALDI-TOF) *m/z* Calcd. for $C_{183}H_{210}NaO_{22}$ ($[M+Na]^+$): 2782.52. Found: 2782.10. Anal. ($C_{183}H_{210}O_{22}$) Calcd: C 79.59, H 7.66. Found: C 79.31, H 7.74. Mp 85–87 °C (CH_2Cl_2-hexane). $[\alpha]^{27}_D$ 23.3 (c = 1.0, $CHCl_3$). IR (KBr) 2925, 2853, 1718, 1594, 1158 cm^{-1}. ^{1}H NMR (400 MHz, $CDCl_3$) δ 0.86 (15H, t, J = 6.8 Hz), 1.25–1.32 (46H, m), 1.41 (10H, br), 1.63 (16H, m), 1.75 (10H, quint, J = 7.6 Hz), 1.85 (10H, quint, J = 4.4 Hz), 1.90 (24H, s), 3.59 (2H, d, J = 11.6 Hz), 3.88 (2H, t, J = 8.8 Hz), 4.30 (10H, t, J = 6.4 Hz), 5.45 (2H, t, J = 2.8 Hz), 5.64 (16H, q, J = 12.0 Hz), 7.02 (2H, s), 7.04 (3H, s), 7.40 (2H, s), 7.44 (8H, d, J = 7.2 Hz), 7.46 (3H, s), 7.50 (5H, s), 7.62 (8H, t, J = 7.6 Hz), 7.94 (8H, s), 8.05 (8H, d, J = 8.0 Hz). ^{13}C NMR (100 MHz, $CDCl_3$), δ 14.1, 18.7, 22.6, 23.5, 25.1, 26.0, 28.7, 29.3, 29.5, 30.2, 31.9, 62.1, 65.3, 65.4, 68.9, 69.0, 96.5, 107.1, 108.5, 108.70, 108.74, 110.8, 120.9, 126.0, 126.1, 126.3, 126.47, 126.53, 128.4, 130.65, 130.68, 131.00, 131.03, 131.07, 131.09, 131.2, 131.7, 132.5, 132.7, 137.0, 137.1, 158.1, 159.8, 160.0, 166.35, 166.37.

Deprotected tetramer (*M*)-Ox-4H. Under an argon atmosphere, to a solution of (*M*)-Ox-**4** (84.5 mg, 30.6 μmol) in dichloromethane/methanol (1:1, 9.5 mL), a catalytic amount of *p*-toluenesulfonic acid monohydrate (11.6 mg, 61.2 μmol) was added, and the mixture was stirred at room temperature for 12 h. The reaction was quenched by adding water, and the organic materials were extracted with dichloromethane three times. The combined organic layers were washed with water, brine, and dried over sodium sulfate. The solvent was evaporated under reduced pressure, and purification by silica gel chromatography (hexane:ethyl acetate = 4:1 and 2:1) and GPC (toluene) gave (*M*)-Ox-**4H** (60.5 mg, 23.3 μmol, 76%). MS (MALDI-TOF) *m/z* Calcd. for $C_{173}H_{194}NaO_{20}$ ($[M+Na]^+$): 2614.41. Found: 2613.67. Anal. ($C_{173}H_{194}O_{20}$) Calcd: C 80.12, H 7.54. Found: C 79.97, H 7.67. Mp 103–105 °C (CH_2Cl_2-hexane). $[\alpha]^{27}_D$ 32.0 (c = 0.1, $CHCl_3$). IR (KBr) 3420, 2925, 2853, 1717, 1595, 1157 cm^{-1}. ^{1}H NMR (400 MHz, $CDCl_3$) δ 0.84 (15H, t, J = 6.8 Hz), 1.23–1.31

(60H, m), 1.39 (10H, br), 1.72 (10H, quint, $J = 7.2$ Hz), 1.93 (24H, s), 4.30 (10H, t, $J = 6.8$ Hz), 5.58 (16H, q, $J = 12.8$ Hz), 5.84 (2H, s), 6.78 (2H, s), 7.01 (3H, s), 7.20 (2H, s), 7.37 (2H, s), 7.39 (8H, d, $J = 10.0$ Hz), 7.48 (6H, s), 7.57 (8H, t, $J = 7.6$ Hz), 7.86 (8H, s), 8.00 (8H, d, $J = 8.4$ Hz). ^{13}C NMR (100 MHz, $CDCl_3$), δ 14.1, 22.6, 23.5, 26.0, 28.6, 28.7, 29.3, 29.50, 29.52, 31.9, 65.4, 65.5, 68.8, 68.9, 107.2, 107.3, 108.1, 108.7, 109.7, 120.9, 125.9, 126.1, 126.3, 126.38, 126.44, 128.4, 130.57, 130.61, 130.96, 130.98, 131.04, 131.6, 132.58, 132.62, 137.01, 137.03, 157.0, 159.9, 160.0, 166.4, 166.6, 166.7.

Pentamer (*P*)-Ox-5. Under an argon atmosphere, to (*P*)-Ox-**3H** (103.8 mg, 51.4 μmol), a solution of (*P*)-**10** (125.5 mg, 0.170 mmol) in *N,N*-dimethylformamide (14.5 mL) and potassium carbonate (28.4 mg, 0.205 mmol) was added at 0 °C, and the mixture was stirred at 35 °C for 4 h. The reaction was quenched by adding water, and the organic materials were extracted with ethyl acetate three times. The combined organic layers were washed with water and brine, and dried over sodium sulfate. The solvent was evaporated under reduced pressure, and purification by silica gel chromatography (hexane:ethyl acetate = 8:1 and 4:1) and GPC (toluene) gave (*P*)-Ox-**5** (155.2 mg, 46.5 μmol, 90%). MS (MALDI-TOF) *m/z* Calcd. for $C_{222}H_{252}NaO_{26}$ ([M+Na]$^+$): 3356.83. Found: 3356.56. Anal. ($C_{222}H_{252}O_{26}$) Calcd: C 79.92, H 7.61. Found: C 79.86, H 7.65. Mp 86–88 °C (CH_2Cl_2-hexane). $[\alpha]^{27}{}_D$ −27.6 (c = 0.1, $CHCl_3$). IR (KBr) 2925, 2854, 1718, 1594, 1158 cm^{-1}. ^{1}H NMR (400 MHz, $CDCl_3$) δ 0.86 (18H, t, $J = 6.8$ Hz), 1.24–1.32 (54H, m), 1.41 (12H, br), 1.63 (18H, m), 1.75 (12H, quint, $J = 7.6$ Hz), 1.85 (12H, quint, $J = 4.4$ Hz), 1.95 (30H, s), 3.59 (2H, d, $J = 11.6$ Hz), 3.88 (2H, t, $J = 8.8$ Hz), 4.30 (12H, t, $J = 6.4$ Hz), 5.45 (2H, t, $J = 2.8$ Hz), 5.63 (20H, q, $J = 12.0$ Hz), 7.02 (2H, s), 7.04 (4H, s), 7.40 (2H, s), 7.44 (10H, d, $J = 7.2$ Hz), 7.46 (4H, s), 7.50 (6H, s), 7.62 (10H, t, $J = 7.6$ Hz), 7.92 (10H, s), 8.02 (10H, d, $J = 8.0$ Hz). ^{13}C NMR (100 MHz, $CDCl_3$), δ 14.1, 18.7, 22.6, 23.5, 25.1, 26.0, 28.7, 29.3, 29.5, 30.2, 31.8, 62.1, 65.3, 65.4, 68.9, 69.0, 96.5, 107.1, 108.5, 108.7, 110.8, 120.9, 126.0, 126.1, 126.3, 126.47, 126.53, 128.4, 130.7, 131.0, 131.07, 131.09, 131.2, 131.6, 132.5, 132.7, 137.0, 158.1, 159.8, 160.0, 166.4.

Deprotected pentamer (*P*)-Ox-5H. Under an argon atmosphere, to a solution of (*P*)-Ox-**5** (69.7 mg, 20.9 μmol) in dichloromethane/methanol (1:1, 5.7 mL), a catalytic amount of *p*-toluenesulfonic acid monohydrate (8.0 mg, 41.8 μmol) was added, and the mixture was stirred at room temperature for 3 h. The reaction was quenched by adding water, and the organic materials were extracted with dichloromethane three times. The combined organic layers were washed with water, brine, and dried over sodium sulfate. The solvent was evaporated under reduced pressure, and purification by silica gel chromatography (hexane:ethyl acetate = 4:1 and 2:1) and GPC (toluene) gave (*P*)-Ox-**5H** (62.2 mg, 19.6 μmol, 94%). MS (MALDI-TOF) *m/z* Calcd. for $C_{212}H_{236}NaO_{24}$ ([M+Na]$^+$): 3188.71. Found: 3188.49. Anal. ($C_{212}H_{236}O_{24}$) Calcd: C 80.37, H 7.51. Found: C 80.17, H 7.52. Mp 101–103 °C (CH_2Cl_2-hexane). $[\alpha]^{27}{}_D$ −32.7 (c = 1.0, $CHCl_3$). IR (KBr) 3420, 2952, 2924, 2853, 1718, 1595, 1157 cm^{-1}. ^{1}H NMR (400 MHz, $CDCl_3$) δ 0.83 (18H, t, $J = 6.8$ Hz), 1.23–1.34 (72H, m), 1.38 (12H, br), 1.73 (12H, quint, $J = 7.2$ Hz), 1.91 (30H, s), 4.29 (12H, t, $J =$

6.8 Hz), 5.54 (20H, q, J = 12.8 Hz), 6.06 (2H, s), 6.78 (2H, s), 7.00 (4H, s), 7.22 (2H, s), 7.36 (2H, s), 7.38 (10H, d, J = 10.0 Hz), 7.48 (8H, s), 7.55 (10H, t, J = 7.6 Hz), 7.83 (10H, s), 7.98 (10H, d, J = 8.4 Hz). ^{13}C NMR (100 MHz, $CDCl_3$), δ 14.1, 22.6, 23.5, 26.0, 28.6, 28.7, 29.3, 29.49, 29.52, 31.8, 65.4, 65.5, 65.6, 68.8, 68.9, 107.16, 107.22, 108.0, 108.7, 109.7, 120.9, 125.9, 126.0, 126.3, 126.37, 126.43, 126.5, 128.4, 130.56, 130.61, 130.96, 130.98, 131.03, 131.6, 132.58, 132.61, 132.64, 137.00, 137.01, 157.1, 159.91, 159.93, 160.0, 166.4, 166.5, 166.6.

Hexamer (*M*)-Ox-6. Under an argon atmosphere, to (*M*)-Ox-**4H** (328.9 mg, 0.127 mmol), a solution of (*M*)-**10** (469.1 mg, 0.634 mmol) in *N,N*-dimethylformamide (38.1 mL) and potassium carbonate (105.2 mg, 0.761 mmol) was added at 0 °C, and the mixture was stirred at 35 °C for 3 h. The reaction was quenched by adding water, and the organic materials were extracted with ethyl acetate three times. The combined organic layers were washed with water and brine, and dried over sodium sulfate. The solvent was evaporated under reduced pressure, and purification by silica gel chromatography (hexane:ethyl acetate = 8:1 and 4:1) and GPC (toluene) gave (*M*)-Ox-**6** (397.8 mg, 0.102 mmol, 80%). MS (MALDI-TOF) *m/z* Calcd. for $C_{261}H_{294}NaO_{30}$ ([M+Na]$^+$): 3931.14. Found: 3931.14. Anal. ($C_{261}H_{294}O_{30}$) Calcd: C 80.15, H 7.58. Found: C 80.12, H 7.68. Mp 90–92 °C (CH_2Cl_2-hexane). $[\alpha]^{27}{}_D$ 22.5 (c = 0.1, $CHCl_3$), IR (KBr) 2925, 2853, 1718, 1594, 1158 cm^{-1}. ^{1}H NMR (400 MHz, $CDCl_3$) δ 0.84 (21H, t, J = 6.8 Hz), 1.24–1.32 (62H, m), 1.41 (14H, br), 1.63 (20H, m), 1.74 (14H, quint, J = 7.6 Hz), 1.85 (14H, quint, J = 4.4 Hz), 1.95 (36H, s), 3.60 (2H, d, J = 11.6 Hz), 3.88 (2H, t, J = 8.8 Hz), 4.31 (14H, t, J = 6.4 Hz), 5.46 (2H, t, J = 2.8 Hz), 5.62 (24H, q, J = 12.0 Hz), 7.02 (2H, s), 7.04 (5H, s), 7.39 (2H, s), 7.44 (12H, d, J = 7.2 Hz), 7.46 (5H, s), 7.50 (7H, s), 7.57 (12H, t, J = 7.6 Hz), 7.91 (12H, s), 8.02 (12H, d, J = 8.0 Hz). ^{13}C NMR (100 MHz, $CDCl_3$), δ 14.1, 18.7, 22.6, 23.5, 25.1, 26.0, 28.7, 29.3, 29.5, 30.2, 31.8, 62.1, 65.3, 65.4, 68.9, 69.0, 96.5, 107.1, 108.5, 108.7, 110.8, 120.9, 126.0, 126.1, 126.3, 126.46, 126.51, 128.4, 130.7, 131.0, 131.06, 131.08, 131.14, 131.6, 132.5, 132.7, 137.0, 158.1, 159.8, 160.0, 166.4.

Deprotected tetramer (*M*)-Ox-6H. Under an argon atmosphere, to a solution of (*M*)-Ox-**4** (84.5 mg, 30.6 μmol) in dichloromethane/methanol (1:1, 9.5 mL), a catalytic amount of *p*-toluenesulfonic acid monohydrate (11.6 mg, 61.2 μmol) was added, and the mixture was stirred at room temperature for 12 h. The reaction was quenched by adding water, and the organic materials were extracted with dichloromethane three times. The combined organic layers were washed with water, brine, and dried over sodium sulfate. The solvent was evaporated under reduced pressure, and purification by silica gel chromatography (hexane:ethyl acetate = 4:1 and 2:1) and GPC (toluene) gave (*M*)-Ox-**4H** (60.5 mg, 23.3 μmol, 76%). MS (MALDI-TOF) *m/z* Calcd for $C_{173}H_{194}NaO_{20}$ ([M+Na]$^+$): 2614.41. Found: 2613.67. Anal. ($C_{173}H_{194}O_{20}$) Calcd. for C 80.12, H 7.54. Found: C 79.97, H 7.67. Mp 103–105 °C (CH_2Cl_2-hexane). $[\alpha]^{27}{}_D$ 32.0 (c = 0.1, $CHCl_3$). IR (KBr) 3420, 2925, 2853, 1717, 1595, 1157 cm^{-1}. ^{1}H NMR (400 MHz, $CDCl_3$) δ 0.84 (15H, t, J = 6.8 Hz), 1.23–1.31 (60H, m), 1.39 (10H, br), 1.72 (10H, quint, J = 7.2 Hz), 1.93 (24H, s), 4.30 (10H, t, J = 6.8 Hz), 5.58 (16H, q, J = 12.8 Hz), 5.84 (2H, s), 6.78 (2H, s), 7.01 (3H, s),

7.20 (2H, s), 7.37 (2H, s), 7.39 (8H, d, J = 10.0 Hz), 7.48 (6H, s), 7.57 (8H, t, J = 7.6 Hz), 7.86 (8H, s), 8.00 (8H, d, J = 8.4 Hz). ^{13}C NMR (100 MHz, $CDCl_3$), δ 14.1, 22.6, 23.5, 26.0, 28.6, 28.7, 29.3, 29.50, 29.52, 31.9, 65.4, 65.5, 68.8, 68.9, 107.2, 107.3, 108.1, 108.7, 109.7, 120.9, 125.9, 126.1, 126.3, 126.38, 126.44, 128.4, 130.57, 130.61, 130.96, 130.98, 131.04, 131.6, 132.58, 132.62, 137.01, 137.03, 157.0, 159.9, 160.0, 166.4, 166.6, 166.7.

Heptamer (*P*)-Ox-7. Under an argon atmosphere, to (*P*)-Ox-**5H** (39.1 mg, 12.3 μmol), a solution of (*P*)-**10** (45.7 mg, 61.7 μmol) in *N,N*-dimethylformamide (3.7 mL) and potassium carbonate (10.2 mg, 74.0 μmol) was added at 0 °C, and the mixture was stirred at 35 °C for 2 h. The reaction was quenched by adding water, and the organic materials were extracted with ethyl acetate three times. The combined organic layers were washed with water and brine, and dried over sodium sulfate. The solvent was evaporated under reduced pressure, and purification by silica gel chromatography (hexane:ethyl acetate = 8:1 and 4:1) and GPC (toluene) gave (*P*)-Ox-**7** (43.0 mg, 9.6 μmol, 78%). MS (MALDI-TOF) *m/z* Calcd. for $C_{300}H_{336}NaO_{34}$ ([M+Na]$^+$): 4505.45. Found: 4505.38. Anal. ($C_{300}H_{336}O_{34}$) Calcd: C 80.32, H 7.55. Found: C 80.15, H 7.35. Mp 95–96 °C (CH_2Cl_2-hexane). $[\alpha]^{27}{}_D$ −23.6 (c = 1.0, $CHCl_3$). IR (KBr) 2925, 2854, 1718, 1594, 1157 cm^{-1}. ^{1}H NMR (400 MHz, $CDCl_3$) δ 0.83 (24H, t, J = 6.8 Hz), 1.24–1.33 (68H, m), 1.41 (16H, br), 1.63 (24H, m), 1.74 (16H, quint, J = 7.6 Hz), 1.84 (16H, quint, J = 4.4 Hz), 1.94 (42H, s), 3.58 (2H, d, J = 11.6 Hz), 3.87 (2H, t, J = 8.8 Hz), 4.31 (16H, t, J = 6.4 Hz), 5.44 (2H, t, J = 2.8 Hz), 5.60 (28H, q, J = 12.0 Hz), 7.03 (8H, s), 7.40 (2H, s), 7.41 (14H, d, J = 7.2 Hz), 7.46 (2H, s), 7.50 (12H, s), 7.56 (14H, t, J = 7.6 Hz), 7.89 (14H, s), 8.01 (14H, d, J = 8.0 Hz). ^{13}C NMR (100 MHz, $CDCl_3$), δ 14.1, 18.7, 22.6, 23.5, 25.1, 26.0, 28.7, 29.3, 29.5, 30.2, 31.9, 62.1, 65.3, 65.4, 68.9, 69.0, 96.5, 107.1, 108.5, 108.7, 110.8, 120.9, 126.0, 126.1, 126.3, 126.47, 126.52, 128.4, 130.7, 131.0, 131.1, 131.2, 131.6, 132.5, 132.7, 137.1, 158.1, 159.8, 160.0, 166.4.

Deprotected heptamer (*P*)-Ox-7H. Under an argon atmosphere, to a solution of (*P*)-Ox-**3** (43.0 mg, 9.6 μmol) in dichloromethane/methanol (1:1, 3.0 mL), a catalytic amount of *p*-toluenesulfonic acid monohydrate (3.7 mg, 19.2 μmol) was added, and the mixture was stirred at room temperature for 6 h. The reaction was quenched by adding water, and the organic materials were extracted with dichloromethane three times. The combined organic layers were washed with water, brine, and dried over sodium sulfate. The solvent was evaporated under reduced pressure, and purification by silica gel chromatography (hexane:ethyl acetate = 4:1 and 2:1) and GPC (toluene) gave (*P*)-Ox-**7H** (41.4 mg, 9.5 μmol, 99%). MS (MALDI-TOF) *m/z* Calcd. for $C_{290}H_{320}NaO_{32}$ ([M+Na]$^+$): 4337.33. Found: 4336.99. Anal. ($C_{290}H_{320}O_{32}$) Calcd: C 80.67, H 7.47. Found: C 80.49, H 7.61. Mp 108–110 °C (CH_2Cl_2-hexane). $[\alpha]^{27}{}_D$ −28.4 (c = 1.0, $CHCl_3$). IR (KBr) 3364, 2953, 2924, 2853, 1718, 1595, 1158 cm^{-1}. ^{1}H NMR (400 MHz, $CDCl_3$) δ 0.82 (24H, t, J = 6.8 Hz), 1.23–1.31 (96H, m), 1.39 (16H, br), 1.73 (16H, quint, J = 7.2 Hz), 1.91 (42H, s), 4.29 (16H, t, J = 6.8 Hz), 5.56 (28H, q, J = 12.8 Hz), 5.80 (2H, s), 6.78 (2H, s), 7.01 (6H, s), 7.19 (2H, s), 7.36 (2H, s), 7.38 (14H, d, J = 10.0 Hz), 7.48 (12H, s), 7.55 (14H, t, J =

7.6 Hz), 7.83 (14H, s), 7.99 (14H, d, $J = 8.4$ Hz). ^{13}C NMR (100 MHz, $CDCl_3$), δ 14.1, 22.6, 23.5, 26.0, 28.7, 29.3, 29.49, 29.51, 31.8, 65.4, 65.48, 65.52, 68.9, 107.1, 107.2, 108.0, 108.7, 109.7, 120.7, 125.9, 126.04, 126.1, 126.3, 126.36, 126.42, 126.5, 128.4, 130.6, 130.97, 131.02, 131.6, 132.6, 132.7, 137.0, 157.1, 159.9, 160.0, 166.3, 166.45, 166.53.

Nonamer (*P*)-Ox-9. Under an argon atmosphere, to (*P*)-Ox-**7H** (125.3 mg, 29.0 μmol), a solution of (*P*)-**10** (145.4 mg, 0.197 mmol) in *N,N*-dimethylformamide (8.7 mL) and potassium carbonate (32.5 mg, 0.235 mmol) was added at 0 °C, and the mixture was stirred at 35 °C for 7 h. The reaction was quenched by adding water, and the organic materials were extracted with ethyl acetate three times. The combined organic layers were washed with water and brine, and dried over sodium sulfate. The solvent was evaporated under reduced pressure, and purification by silica gel chromatography (hexane:ethyl acetate = 8:1 and 4:1) and GPC (toluene) gave (*P*)-Ox-**9** (137.0 mg, 24.3 μmol, 84%). MS (MALDI-TOF) *m/z* Calcd. for $C_{378}H_{420}NaO_{42}$ ([M+Na]$^+$): 5654.06. Found: 5654.11. Anal. ($C_{378}H_{420}O_{42}$) Calcd: C 80.56, H 7.51. Found: C 80.33, H 7.60. Mp 101–102 °C (CH_2Cl_2-hexane). $[\alpha]^{27}{}_D$ −36.4 (c = 0.1, $CHCl_3$), IR (KBr) 2925, 2853, 1719, 1594, 1157 cm^{-1}. ^{1}H NMR (400 MHz, $CDCl_3$) δ 0.82 (30H, t, $J = 6.8$ Hz), 1.23–1.33 (80H, m), 1.41 (20H, br), 1.63 (30H, m), 1.74 (20H, quint, $J = 7.6$ Hz), 1.85 (20H, quint, $J = 4.4$ Hz), 1.94 (54H, s), 3.58 (2H, d, $J = 11.6$ Hz), 3.87 (2H, t, $J = 8.8$ Hz), 4.30 (20H, t, $J = 6.4$ Hz), 5.44 (2H, t, $J = 2.8$ Hz), 5.60 (36H, q, $J = 12.0$ Hz), 7.03 (10H, s), 7.39 (2H, s), 7.42 (18H, d, $J = 7.2$ Hz), 7.46 (4H, s), 7.50 (16H, s), 7.56 (18H, t, $J = 7.6$ Hz), 7.90 (18H, s), 8.01 (18H, d, $J = 8.0$ Hz). ^{13}C NMR (100 MHz, $CDCl_3$), δ 14.1, 18.7, 22.6, 23.5, 25.1, 26.0, 28.7, 29.3, 29.50, 29.52, 30.2, 31.8, 62.1, 65.3, 65.4, 69.0, 96.5, 107.1, 108.5, 108.7, 110.8, 120.9, 126.1, 126.3, 126.47, 126.52, 128.4, 130.7, 131.0, 131.1, 131.2, 131.6, 132.5, 132.7, 137.0, 158.1, 159.8, 160.0, 166.4.

7.3 Experimental Method for Chap. 3

7.3.1 *CD and UV-Vis Analyses of* (P)*-Ox-5/*(M)*-Ox-6 in Trifluoromethylbenzene*

A solution of (*P*)-Ox-**5**/(*M*)-Ox-**6** (1:1) (total concentration, 0.5 mM) was prepared by dissolving (*P*)-Ox-**5** (0.84 mg, 2.5×10^{-4} mmol) and (*M*)-Ox-**6** (0.98 mg, 2.5×10^{-4} mmol) in trifluoromethylbenzene (1.0 mL) with heating at 80 °C for 3 min. A portion of the solution (0.80 mL) was transferred into a quarz cell (optical path length 0.0107 cm), heated at 60 °C for 20 min, cooled to 5 °C, and allowed to settle for 180 min, during which CD and UV-vis analyses were conducted.

7.3.2 DLS Analysis

A solution of (*P*)-Ox-**5**/(*M*)-Ox-**6** (1:1) (total concentration, 0.5 mM) was prepared by dissolving (*P*)-Ox-**5** (1.25 mg, 3.7×10^{-4} mmol) and (*M*)-Ox-**6** (1.47 mg, 3.8×10^{-4} mmol) in trifluoromethylbenzene (1.5 mL) with heating at 80 °C for 3 min. After transferred to the DLS cell, the solution was heated to 60 °C for 90 min, and cooled to 25 °C for 170 min. The average diameters of 0.7 and 1.6 nm were obtained at 60 and 25 °C, respectively.

7.3.3 Job Plots Experiment

(*P*)-Ox-**5**/(*M*)-Ox-**6** in trifluoromethylbenzene (0.8 mL; total concentration, 0.5 mM) was used. A 1:5 solution was prepared from (*P*)-Ox-**5** (1.11 mg, 3.3×10^{-4} mmol) and (*M*)-Ox-**6** (0.26 mg, 6.6×10^{-5} mmol); a 2:1 solution from (*P*)-Ox-**5** (0.89 mg, 2.7×10^{-4} mmol) and (*M*)-Ox-**6** (0.52 mg, 1.3×10^{-4} mmol); a 1:1 solution from (*P*)-Ox-**5** (0.67 mg, 2.0×10^{-4} mmol) and (*M*)-Ox-**6** (0.78 mg, 2.0×10^{-4} mmol); a 1:2 solution from (*P*)-Ox-**5** (0.45 mg, 1.3×10^{-4} mmol) and (*M*)-Ox-**6** (1.04 mg, 2.7×10^{-4} mmol); a 1:5 solution from (*P*)-Ox-**5** (0.22 mg, 6.6×10^{-5} mmol) and (*M*)-Ox-**6** (1.31 mg, 3.3×10^{-4} mmol). The solutions were heated at 60 °C for 20 min in a quartz cell (optical path length 0.0107 cm), cooled to 5 °C, and allowed to settle for 10–12 h, when the CD (316 nm) value changes were less than 1 mdeg within a 10 min period. Then, the solution phase was removed, and the CD values (316 nm) of the film on quartz cell surface were obtained. The results showed 1:1 complex formation of (*P*)-Ox-**5**/(*M*)-Ox-**6**.

7.3.4 Separation Experiment

A trifluoromethylbenzene solution (0.80 mL) of (*P*)-Ox-**5**/(*M*)-Ox-**6** (1:1) (total concentration, 0.5 mM) in a quartz cell **A** (optical path length 0.0107 cm) was heated at 60 °C for 20 min, cooled to 5 °C, and allowed to settle for each time. Then, the solution phase was transferred into another quartz cell **B** (optical path length 0.0114 cm), which was immediately analyzed by CD and UV-vis at 5 °C. The quartz cell **A** (optical path length 0.0107 cm) was washed with cooled trifluoromethylbenzene at 5 °C (0.70 mL) twice, and the cell surface was analyzed by CD and UV-vis at 5 °C. The above solution phase in the quartz cell **B** and trifluoromethylbenzene washings were combined in a flask **X**.

Determine the yield of (*P*)-Ox-**5**/(*M*)-Ox-**6** in fibril film and solution phase: The quartz cell **A** was dried *in vacuo*, and chloroform (0.80 mL) was added to dissolve and dissociate fibril film. Then, the solution in the cell **A** was analyzed by UV-vis at 25 °C. The combined solution in the flask **X** was concentrated, and dried *in vacuo*.

The residue was dissolved in chloroform (0.80 mL), and the solution was analyzed by UV-vis at 25 °C. The yield of (*P*)-Ox-**5**/(*M*)-Ox-**6** UV-vis was determined by a calibration curve.

7.3.5 Calibration Curve

A calibration curve using UV-vis absorbances at 300 nm was obtained for solutions of (*P*)-Ox-**5**/(*M*)-Ox-**6** (1:1) in chloroform (total concentration, 0.5, 0.4, 0.2 and 0.1 mM). The solutions were prepared as follows. (*P*)Ox-**5**/(*M*)-Ox-**6** (1:1) (total concentration, 0.5 mM): (*P*)-Ox-**5** (0.836 mg, 2.5×10^{-4} mmol) and (*M*)Ox-**6** (0.980 mg, 2.5×10^{-4} mmol) were dissolved in chloroform (1.0 mL). (*P*)-Ox-**5**/(*M*)-Ox-**6** (1:1) (total concentration, 0.4 mM): (*P*)-Ox-**5** (0.670 mg, 2.0×10^{-4} mmol) and (*M*)-Ox-**6** (0.782 mg, 2.0×10^{-4} mmol) were dissolved in chloroform (1.0 mL). (*P*)-Ox-**5**/(*M*)-Ox-**6** (1:1) (total concentration, 0.2 mM): (*P*)-Ox-**5** (0.335 mg, 1.0×10^{-4} mmol) and (*M*)-Ox-**6** (0.391 mg, 1.0×10^{-4} mmol) were dissolved in chloroform (1.0 mL). (*P*)-Ox-**5**/(*M*)-Ox-**6** (1:1) (total concentration, 0.1 mM): (*P*)-Ox-**5** (0.167 mg, 5.0×10^{-5} mmol) and (*M*)-Ox-**6** (0.195 mg, 5.0×10^{-5} mmol) were dissolved in chloroform (1.0 mL). The solution was transferred to a quartz cell (optical path length 0.0107 cm), and UV-vis analyses (300 nm) gave a calibration curve.

7.3.6 AFM Analysis of Fibril Film Formation Process

AFM images were recorded under ambient conditions using a Digital Instrument Multimode Nanoscope IIIa operating in the tapping mode regime. Micro-fabricated silicon cantilever tips (OMCL-AC160TS-C2) were used. A trifluoromethylbenzene solution (1.0 mL) of (*P*)-Ox-**5**/(*M*)-Ox-**6** (1:1) (total concentration, 0.5 mM) in a vial (2.5 cm in opening diameter) was heated at 80 °C (3 min), and cooled to 25 °C for 3 min. A glass plate was immersed into the solution, and allowed to settle at 5 °C for 5, 10, or 15 min. Then, the plate was taken from the solution, washed with trifluoromethylbenzene (5 °C), and dried *in vacuo*.

7.3.7 Film Seeding Experiment

7.3.7.1 Preparation of a Seed of Fibril Film on a Cell Surface

A solution of (*P*)-Ox-**5**/(*M*)-Ox-**6** (1:1) (total concentration, 0.5 mM) in trifluoromethylbenzene (0.80 mL) in a quartz cell **A** (optical path length 0.0107 cm) was heated at 60 °C (20 min), cooled to 5 °C, and allowed to settle for 60 min.

The solution phase was removed from the cell, and the cell was washed with trifluoromethylbenzene cooled at 5 °C (0.70 mL) twice, and dried *in vacuo*.

7.3.7.2 Seeding Experiment

A solution of (*P*)-Ox-**5**/(*M*)-Ox-**6** (1:1) (total concentration, 0.5 mM) in trifluoromethylbenzene (1.20 mL) in a quartz cell **B** (optical path length 0.0114 cm) was heated at 60 °C (20 min) and cooled to 25 °C (60 min), provided a random-coil solution. Then, the solution (0.80 mL) was added into the quartz cell **A** at 25 °C. The mixture was allowed to settle for 600 min, and was analyzed by CD at 316 nm. The same experiments were conducted at 40 and 50 °C.

7.3.8 Capillary Seeding Experiment

7.3.8.1 Preparation of Fibril Film Seed on a Capillary

A solution of (*P*)-Ox-**5**/(*M*)-Ox-**6** (1:1) (total concentration, 0.5 mM; trifluoromethylbenzene) (10 mL) in a vial (2.5 cm in opening diameter) was heated at 80 °C (3 min), and cooled to 25 °C (3 min). A portion of solution (0.54 mL) was transferred to a glass tube (5.5 mm in opening diameter). A capillary with 4 cm length and 1 mm diameter was immersed into the solution at 2 cm from the tip. The solution was cooled at 5 °C for 10 h. Then, the capillary was taken from the solution, washed with cooled trifluoromethylbenzene at 5 °C, and dried *in vacuo*. The residual film on the capillary was washed with chloroform (5 mL) to dissolve (*P*)-Ox-**5**/(*M*)-Ox-**6**, and the solution was analyzed in a quartz cell (optical path length 1 cm) by UV-vis to determine the amount of (*P*)-Ox-**5**/(*M*)-Ox-**6**. The experiments were repeated twice to confirm reproducibility. The amount of (*P*)-Ox-**5**/(*M*)-Ox-**6** on capillaries in the experiments were determined to be 3.9×10^{-3} and 4.7×10^{-3} μmol by UV-vis analysis.

7.3.8.2 Seeding Experiment Using a Fibril Film Seed on a Capillary

A fibril film seed on a capillary was prepared as above. The capillary was put into a (*P*)-Ox-**5**/(*M*)-Ox-**6** solution (1:1) (total concentration, 0.5 mM; trifluoromethylbenzene) (4.2 mL) in a quartz cell (1 cm optical path length) at 25 °C, in which the capillary was not attached to the wall of the cell. The mixture was allowed to settle at the temperature for 24 h.

Determine the amount of (*P*)-Ox-**5**/(*M*)-Ox-**6** on the capillary surface: The capillary was taken from the solution, washed with cooled trifluoromethylbenzene at 5 °C (1 mL), and dried *in vacuo*. The film on the capillary was washed with chloroform

(50 mL) to dissolve (*P*)-Ox-**5**/(*M*)-Ox-**6**, and the solution was analyzed in a quartz cell (optical path length 1 cm) by UV-vis to determine the amount of (*P*)-Ox-**5**/(*M*)-Ox-**6**.

Determine the amount of (*P*)-Ox-**5**/(*M*)-Ox-**6** on the cell surface: The solution phase in the cell was transferred into a flask **X**. The cell was washed with cooled trifluoromethylbenzene at 5 °C (0.70 mL) twice, and the washings were also transferred into a flask X. The cell was dried *in vacuo*, and chloroform (3 mL) was added to dissolve (*P*)-Ox-**5**/(*M*)-Ox-**6**, and the solution was analyzed in a quartz cell (optical path length 1 cm) by UV-vis to determine the amount of (*P*)-Ox-**5**/(*M*)-Ox-**6**.

Determine the amount of (*P*)-Ox-**5**/(*M*)-Ox-**6** in the solution phase: The above combined solutions in the quartz cell **X** were dried *in vacuo*, and chloroform (10 mL) was added to dissolve (*P*)-Ox-**5**/(*M*)-Ox-**6**. A portion of the solution (1 mL) in chloroform was taken into a volumetric flask (100 mL), to which chloroform was added to make (*P*)-Ox-**5**/(*M*)-Ox-**6** solution (100 mL). The solution was analyzed in a quartz cell (optical path length 1 cm) by UV-vis to determine the amount of (*P*)-Ox-**5**/(*M*)-Ox-**6**.

7.3.9 Surface Effect in the Fibril Films Formation

7.3.9.1 To Determine the Yields of (*P*)-Ox-5/(*M*)-Ox-6 in Fibril Film

Polyethylene terephthalate, polypropylene, cellulose, polyethylene, Teflon, polylactic acid, and aluminium foil: A trifluoromethylbenzene solution (2.0 mL) of (*P*)-Ox-**5**/(*M*)-Ox-**6** (1:1) (total concentration, 0.5 mM) in a vial (1.5 cm in opening diameter) was heated at 80 °C (3 min), and cooled to 25 °C for 3 min. A plate (height 2.5 cm, width 1.0 cm) was immersed in the solution. The solution was cooled to 5 °C, and allowed to settle at 5 °C for 60 min. Then, the plate was taken from the solution, washed with cooled trifluoromethylbenzene at 5 °C (5 mL), dried *in vacuo*, and cut into a piece with 1 cm square. The plate was washed with chloroform (1 mL) to dissolve (*P*)-Ox-**5**/(*M*)-Ox-**6**, and the solution was analyzed in a quartz cell (optical path length 0.1 cm) by UV-vis to determine the yields. The experiments were repeated three times to confirm reproducibility.

Glass or quartz plate: A trifluoromethylbenzene solution (2.0 mL) of (*P*)-Ox-**5**/(*M*)-Ox-**6** (1:1) (total concentration, 0.5 mM) in a vial (1.5 cm in opening diameter) was heated at 80 °C (3 min), and cooled to 25 °C (3 min). A glass or quartz plate (height 2.5 cm, width 1.0 cm, thickness 0.1 cm) was immersed in the solution. The solution was cooled to 5 °C, and allowed to settle at 5 °C for 60 min. Then, the plate was taken from the solution, washed with cooled trifluoromethylbenzene at 5 °C (5 mL), and dried *in vacuo*. A 1 cm square area of the fibril film was obtained by removing the other part by wiping with a cotton bud soaked in chloroform. The residual film was washed with chloroform (1 mL) to dissolve (*P*)-Ox-**5**/(*M*)-Ox-**6**, and the solution was analyzed in a quartz cell (optical path length 0.1 cm) by UV-vis to determine the yields. The experiments were repeated three times to confirm reproducibility.

Glass plate with one side coated with Au: A trifluoromethylbenzene solution (2.0 mL) of (*P*)-Ox-**5**/(*M*)-Ox-**6** (1:1) (total concentration, 0.5 mM) in a vial (1.5 cm in opening diameter) was heated at 80 °C (3 min), and cooled to 25 °C (3 min). A glass with one side coated by Au (height 2.5 cm, width 1.0 cm, depth 0.1 cm) was immersed into the solution. The solution was cooled to 5 °C, and allowed to settle at 5 °C for 60 min. Then, the plate was taken from the solution, washed with cooled trifluoromethylbenzene at 5 °C (5 mL), and dried *in vacuo*. A 1 cm square area of the fibril film was obtained by removing the other part by wiping with a cotton bud soaked in chloroform. The residual film on the Au-coated side was washed with chloroform (1 mL) to dissolve (*P*)-Ox-**5**/(*M*)-Ox-**6**, and the solution was analyzed in a quartz cell (optical path length 0.1 cm) by UV-vis to determine the yields. The experiments were repeated three times to confirm reproducibility.

7.4 Experimental Method for Chap. 4

7.4.1 Stirring Induced Hetero-aggregate Formation and Self-assembly

7.4.1.1 CD and UV-Vis Analysis

A trifluoromethylbenzene solution (5.0 mL) of (*P*)-Ox-**5**/(*M*)-Ox-**6** (1:1) (total concentration, 0.5 mM) was prepared by dissolving (*P*)-Ox-**5** (4.17 mg, 1.25×10^{-3} mmol) and (*M*)-Ox-**6** (4.89 mg, 1.25×10^{-3} mmol) in a cylindrical glass vial (diameter, 21 mm), and was heated at 80 °C for 3 min. Then, the solution was cooled to 25 °C, and was mechanically stirred in clockwise direction with an oval-shaped Teflon magnetic stirring bar (weight, 0.40 g) at the rate of 2000 rpm. After mechanical stirring for a certain period of time, a part of the solution was taken, and analyzed by CD and UV-vis. The CD sample was returned to the reaction mixture, and stirring was continued.

7.4.1.2 Analysis of Vessel Wall Adsorption

A trifluoromethylbenzene solution (2.0 mL) of (*P*)-Ox-**5**/(*M*)-Ox-**6** (1:1) (total concentration, 0.5 mM) was prepared by dissolving (*P*)-Ox-**5** (1.66 mg, 5.0×10^{-4} mmol) and (*M*)-Ox-**6** (1.96 mg, 5.0×10^{-4} mmol) in a cylindrical glass vial (diameter, 12 mm), and was heated at 80 °C for 3 min. Then, the solution was cooled to 25 °C, and was mechanically stirred in clockwise direction with an oval-shaped Teflon magnetic stirring bar (weight, 0.40 g) at the rate of 2000 rpm for 48 h. The solution phase was removed, and the vessel was washed with cooled trifluoromethylbenzene at 0 °C (1.0 mL) 3 times. The vessel was dried *in vacuo*, and chloroform (2.0 mL) was added to dissolve and dissociate fibril film. Essentially no (*P*)-Ox-**5**/(*M*)-Ox-**6** was detected by UV analysis, $\varepsilon < 10^3$ cm^{-1} M^{-1}.

7.4.1.3 Filtration Experiment

A trifluoromethylbenzene solution (5.0 mL) of (*P*)-Ox-**5**/(*M*)-Ox-**6** (1:1) (total concentration, 0.5 mM) was prepared by dissolving (*P*)-Ox-**5** (4.17 mg, 1.25×10^{-3} mmol) and (*M*)-Ox-**6** (4.89 mg, 1.25×10^{-3} mmol) in a cylindrical glass vial (diameter, 21 mm), and was heated at 80 °C for 3 min. Then, the solution was cooled to 25 °C, and was mechanically stirred in clockwise direction with an oval-shaped Teflon magnetic stirring bar (weight, 0.40 g) at the rate of 2000 rpm. After mechanical stirring for a certain period of time, the solution was filtered through 1-μm-pore or 5-μm-pore membrane filter, and the solution was analyzed by CD, UV-vis, and AFM.

7.4.2 High Pressure Experiment

A trifluoromethylbenzene solution (10.0 mL) of (*P*)-Ox-**5**/(*M*)-Ox-**6** (1:1) (total concentration, 0.5 mM) was prepared by dissolving (*P*)-Ox-**5** (8.34 mg, 2.5×10^{-3} mmol) and (*M*)-Ox-**6** (9.79 mg, 2.5×10^{-3} mmol), and was heated at 80 °C for 3 min. Then, the solution was cooled to 25 °C, and was subjected to a high pressure 0.2 GPa for 24 h in polyethylene vessel.

7.4.3 Centrifuge Experiment

A trifluoromethylbenzene solution (2.0 mL) of (*P*)-Ox-**5**/(*M*)-Ox-**6** (1:1) (total concentration, 0.5 mM) was prepared by dissolving (*P*)-Ox-**5** (1.67 mg, 5.0×10^{-4} mmol) and (*M*)-Ox-**6** (1.95 mg, 5.0×10^{-4} mmol) in a cylindrical glass vial (diameter, 12 mm), and was heated at 80 °C for 3 min. Then, the solution was cooled to 25 °C, and centrifuged at 10,000 rpm for 5 h.

7.4.4 Sonication Experiment

A trifluoromethylbenzene solution (2.0 mL) of (*P*)-Ox-**5**/(*M*)-Ox-**6** (1:1) (total concentration, 0.5 mM) was prepared by dissolving (*P*)-Ox-**5** (1.67 mg, 5.0×10^{-4} mmol) and (*M*)-Ox-**6** (1.96 mg, 5.0×10^{-4} mmol) in a cylindorical glass vial (diameter, 12 mm), and was heated at 80 °C for 3 min. Then, the solution was cooled to 25 °C, and was sonicated by 200 W instrument.

7.4.5 *Conversion of Random-Coil Aggregates to Hetero-aggregates by the Mechanical Stirring*

A trifluoromethylbenzene solution (2.0 mL) of (*P*)-Ox-**5**/(*M*)-Ox-**6** (1:1) (total concentration, 0.5 mM) was prepared by dissolving (*P*)-Ox-**5** (1.66 mg, 5.0×10^{-4} mmol) and (*M*)-Ox-**6** (1.96 mg, 5.0×10^{-4} mmol), and was heated at 80 °C for 3 min. After cooling to 5 °C for 10 h, the solution phase was transferred to another glass vessel (diameter, 12 mm), and was mechanically stirred in clockwise direction with an oval-shaped Teflon magnetic stirring bar (weight, 0.40 g) at the rate of 1500 rpm at 25 °C for 120 h.

7.5 Experimental Method for Chap. 5

Hexamer (*P*)-Ox-6. Anal. ($C_{261}H_{294}O_{30}$) Calcd: H, 7.58; C, 80.15. Found: H, 7.78; C, 80.17.

7.5.1 *Chiral Symmetry Breaking Experiment*

A 50:50 mixture of (*P*)-Ox-**6**/(*M*)-Ox-**6** in trifluoromethylbenzene (total concentration, 0.5 mM) was prepared. (*P*)-Ox-**6** (2.93 mg, 7.5×10^{-4} mmol) and (*M*)-Ox-**6** (2.93 mg, 7.5×10^{-4} mmol) were weighed in a cylindrical glass vial (diameter, 21 mm) by a microbalance within error of 0.5%. The samples were dissolved in trifluoromethylbenzene (3.0 mL, total concentration, 0.5 mM) by heating to 90 °C for 10 min. Then, the solution was cooled to 25 °C, and mechanically stirred in clockwise direction with an oval-shaped Teflon magnetic stirring bar (weight, 0.40 g) at the rate of 2000 rpm at 25 °C. Then, CD and UV-vis spectra were obtained.

7.5.2 *Seeding Experiment*

A 50:50 mixture of (*P*)-Ox-**6**/(*M*)-Ox-**6** in trifluoromethylbenzene (1.0 mL; total concentration, 0.5 mM) was prepared by dissolving (*P*)-Ox-**6** (0.978 mg, 2.5×10^{-4} mmol) and (*M*)-Ox-**6** (0.978 mg, 2.5×10^{-4} mmol) by heating to 90 °C for 10 min in a cylindrical glass vial (diameter, 12 mm). Then, a part of the solution (0.2 mL) was cooled to 25 °C, and was mechanically stirred in clockwise direction with an oval-shaped Teflon magnetic stirring bar (weight, 0.40 g) at the rate of 2000 rpm for 73 h (solution I). Then, the solution I (0.2 mL) was mixed with the rest of solution containing a 50:50 mixture of random-coil (*P*)-Ox-**6**/(*M*)-Ox-**6** in trifluoromethylbenzene (0.8 mL; total concentration, 0.5 mM). The resulted mixture was allowed to settle for 50 min, and CD and UV-vis analyses were conducted.

7.5.3 *Stochastic Chiral Symmetry Breaking by Mixing Experiments*

7.5.3.1 Mixing of Hetero-double-helix B (55:45) and Random-Coil (45:55)

A 55:45 mixture of (*P*)-Ox-**6**/(*M*)-Ox-**6** in trifluoromethylbenzene (1.0 mL; total concentration, 0.5 mM) was prepared by dissolving (*P*)-Ox-**6** (1.08 mg, 2.75 × 10^{-4} mmol) and (*M*)-Ox-**6** (0.880 mg, 2.25 × 10^{-4} mmol) by heating to 90 °C for 10 min in a cylindrical glass vial (diameter, 12 mm). Then, the solution was cooled to 25 °C, and was mechanically stirred in clockwise direction with an oval-shaped Teflon magnetic stirring bar (weight, 0.40 g) at the rate of 2000 rpm for 50 h (solution I). A 45:55 mixture of (*P*)-Ox-**6**/(*M*)-Ox-**6** in trifluoromethylbenzene (1.0 mL; total concentration, 0.5 mM) was prepared by dissolving (*P*)-Ox-**6** (0.880 mg, 2.25 × 10^{-4} mmol) and (*M*)-Ox-**6** (1.08 mg, 2.75 × 10^{-4} mmol) by heating to 90 °C for 10 min in a cylindrical glass vial (diameter, 12 mm) (solution II). A part of solution I (0.5 mL) and a part of solution II (0.5 mL) were mixed, and mechanically stirred in clockwise direction with an oval-shaped Teflon magnetic stirring bar (weight, 0.40 g) at the rate of 2000 rpm for 45 h. Then, CD and UV-vis analyses were conducted.

7.5.3.2 Mixing of Hetero-double-helix Ent-B (45:55) and Random-Coil (55:45)

A 45:55 mixture of (*P*)-Ox-**6**/(*M*)-Ox-**6** in trifluoromethylbenzene (1.0 mL; total concentration, 0.5 mM) was prepared by dissolving (*P*)-Ox-**6** (0.880 mg, 2.25 × 10^{-4} mmol) and (*M*)-Ox-**6** (1.08 mg, 2.75 × 10^{-4} mmol) by heating to 90 °C for 10 min in a cylindrical glass vial (diameter, 12 mm). Then, the solution was cooled to 25 °C, and was mechanically stirred in clockwise direction with an oval-shaped Teflon magnetic stirring bar (weight, 0.40 g) at the rate of 2000 rpm for 70 h (solution I). A 55:45 mixture of (*P*)-Ox-**6**/(*M*)-Ox-**6** in trifluoromethylbenzene (1.0 mL; total concentration, 0.5 mM) was prepared by dissolving (*P*)-Ox-**6** (1.08 mg, 2.75 × 10^{-4} mmol) and (*M*)-Ox-**6** (0.880 mg, 2.25 × 10^{-4} mmol) by heating to 90 °C for 10 min in a cylindrical glass vial (diameter, 12 mm) (solution II). A part of solution I (0.5 mL) and a part of solution II (0.5 mL) were mixed, and mechanically stirred in clockwise direction with an oval-shaped Teflon magnetic stirring bar (weight, 0.40 g) at the rate of 2000 rpm for 22 h. Then, CD UV-vis analyses were conducted.

7.5.4 *Proximate Stochastic Chiral Symmetry Breaking by Mixing Experiments*

7.5.4.1 Mixing of Hetero-double-helix B (55:45) and Homo-double-helix (*M*)-Ox-6

A solution of (*M*)-Ox-**6** in trifluoromethylbenzene (0.50 mL; concentration, 0.5 mM) was prepared by dissolving (*M*)-Ox-**6** (0.978 mg, 2.5×10^{-4} mmol) by heating to 90 °C for 10 min and cooled to 0 °C (solution I). A 55:45 mixture of (*P*)-Ox-**6**/(*M*)-Ox-**6** in trifluoromethylbenzene (1.5 mL; total concentration, 0.5 mM) was prepared by dissolving (*P*)-Ox-**6** (1.61 mg, 4.13×10^{-4} mmol) and (*M*)-Ox-**6** (1.32 mg, 3.38×10^{-4} mmol) by heating to 90 °C for 10 min in a cylindrical glass vial (diameter, 12 mm). Then, the solution was cooled to 25 °C, and was mechanically stirred in clockwise direction with an oval-shaped Teflon magnetic stirring bar (weight, 0.40 g) at the rate of 2000 rpm for 24 h (solution II). A part of solution I (0.1 mL) and a part of solution II (1.0 mL) were mixed, and mechanically stirred in clockwise direction with an oval-shaped Teflon magnetic stirring bar (weight, 0.40 g) at the rate of 2000 rpm for 20 h. Then, CD and UV-vis analyses were conducted.

7.5.4.2 Mixing of Hetero-double-helix B (45:55) and Homo-double-helix (*P*)-Ox-6

A solution of (*P*)-Ox-**6** in trifluoromethylbenzene (0.50 mL; concentration, 0.5 mM) was prepared by dissolving (*P*)-Ox-**6** (0.978 mg, 2.5×10^{-4} mmol) with heating to 90 °C for 10 min and cooled to 0 °C (solution I). A 45:55 mixture of (*P*)-Ox-**6**/(*M*)-Ox-**6** in trifluoromethylbenzene (1.5 mL; total concentration, 0.5 mM) was prepared by dissolving (*P*)-Ox-**6** (1.32 mg, 3.38×10^{-4} mmol) and (*M*)-Ox-**6** (1.61 mg, 4.13×10^{-4} mmol) with heating to 90 °C for 10 min in a cylindrical glass vial (diameter, 12 mm). Then, the solution was cooled to 25 °C, and was mechanically stirred in clockwise direction with an oval-shaped Teflon magnetic stirring bar (weight, 0.40 g) at the rate of 2000 rpm for 25 h (solution II). A part of solution I (0.1 mL) and a part of solution II (1.0 mL) were mixed, and mechanically stirred in clockwise direction with an oval-shaped Teflon magnetic stirring bar (weight, 0.40 g) at the rate of 2000 rpm for 40 h. Then, CD and UV-vis analyses were conducted, in which CD spectrum inverted from that of solution I.

Curriculum Vitae

Tsukasa Sawato

Office Address: Graduate School of Pharmaceutical Sciences, Tohoku University,
6-3 Aoba, Aramaki, Aoba, Sendai 980-8578, Japan
Phone: +81-22-795-6813
FAX: +81-22-795-6811
Email: Tsukasa.sawato.b1@tohoku.ac.jp
Nationality: Japanese

WORK EXPERIMENCE

JSPS (Japan Society for the Promotion of Science) Predoctoral Fellow	2018 Apr–2020 Mar

EDUCATION

Tohoku University, Sendai, Japan Ph.D. in Pharmaceutical Sciences	2016 Apr–2019 Mar
Tohoku University, Sendai, Japan M.S. in Pharmaceutical Sciences	2014 Apr–2016 Mar
Tohoku University, Sendai, Japan B.S. in Pharmaceutical Sciences	2009 Apr–2014 Mar

T. Sawato, *Synthesis of Optically Active Oxymethylenehelicene Oligomers and Self-assembly Phenomena at a Liquid–Solid Interface*, Springer Theses,
https://doi.org/10.1007/978-981-15-3192-7

The manufacturer's authorised representative in the EU is Springer Nature Customer Service Centre GmbH, Europaplatz 3, 69115 Heidelberg, Germany. If you have any concerns regarding our products, please contact ProductSafety@springernature.com

Printed and bound by CPI Group (UK) Ltd, Croydon, CR0 4YY
17/07/2026
02170255-0001